Ravenswood

Ravenswood

THE STEELWORKERS' VICTORY AND
THE REVIVAL OF AMERICAN LABOR

Tom Juravich

and

Kate Bronfenbrenner

ILR Press

an imprint of

Cornell University Press

Ithaca and London

First published 1999 by Cornell University Press
First printing, Cornell Paperbacks, 2000

Printed in the United States of America

LIBRARY OF CONGRESS CATALOGING-IN-PUBLICATION DATA

Juravich, Tom.
    Ravenswood : the steelworkers' victory and the revival of American
labor / Tom Juravich and Kate Bronfenbrenner.
        p. cm.
    Includes index.
    ISBN 0-8014-3633-8 (cloth : alk. paper)
    ISBN 0-8014-8666-1 (pbk. : alk. paper)
    1. Strikes and lockouts—Aluminum industry—West Virginia—
Ravenswood—History. 2. United Steelworkers of America. Local
5668 (Ravenswood, W. Va.)—History. 3. Ravenswood Aluminum Co.—
History. I. Bronfenbrenner, Kate, 1954-. II. Title.
HD5325 .A492 1990.J87 1999
331.89298691429 0975431—dc21   99-17965

Cornell University Press strives to use environmentally responsible suppliers and materials to the fullest extent possible in the publishing of its books. Such materials include vegetable-based, low-VOC inks and acid-free papers that are recycled, totally chlorine-free, or partly composed of nonwood fibers. Books that bear the logo of the FSC (Forest Stewardship Council) use paper taken from forests that have been inspected and certified as meeting the highest standards for environmental and social responsibility. For further information, visit our website at www.cornellpress.cornell.edu.

3 5 7 9   Cloth printing   10 8 6 4 2
1 3 5 7 9   Paperback printing   10 8 6 4 2

FSC FSC Trademark © 1996 Forest Stewardship Council A.C.
      SW-COC-098

For the members of USWA Local 5668,
their families, and their union

Contents

INTRODUCTION

When the Ravenswood Aluminum Company locked out seventeen hundred workers on October 31, 1990, it hardly looked like a big opportunity for labor. In what had become standard operating procedure for employers during the 1980s, management broke off bargaining with the United Steelworkers of America, and then brought hundreds of replacement workers into a heavily fortified plant surrounded by barbed wire and security cameras. Injunctions prevented union members from doing little more than symbolic picketing, and the wheels of justice, as they had done for more than a decade, creaked ever so slowly. All the pieces were in place for another long, drawn-out defeat for labor.

The seventeen hundred employees of the Ravenswood Aluminum Company were not the only workers in the 1980s who felt the ground shift under their jobs. A bold new group of CEOs, financial consultants, and corporate raiders had emerged who abandoned traditional business practices. They quickly went to work buying and selling companies that had been long-term, stable American employers. Using junk bonds and leveraged buyouts, moving money around at lightning speed, they left American workers in a state of shock, watching as their work was downsized, contracted out, sped up, streamlined, and eliminated. Production was global, American workers were told, and they would have to compete in a worldwide market.

Many lost their jobs and the ability to earn a living, while others just lost their will to fight back. Of those who kept their jobs, many were forced to take massive concessions in wages and benefits. Later they watched as the number of hours they worked steadily increased to make up for the concessions and cutbacks. Their job security had disappeared and in its place came new expectations, new responsibilities—multi-skilling, multi-tasking, and "continuous quality improvement." The net result was that jobs were harder, more stressful, and considerably less safe and secure than even half a generation earlier.

During the 1980s, unions were largely defined by their losses. Little that labor did was effective in stemming the tide of corporate power and

the assault on workers' rights and decent jobs. Unions suffered big public defeats at PATCO, Eastern Airlines, International Paper, Hormel,
Phelps-Dodge. Strike activity plummeted to its lowest level since the
1920s.

Labor also lost its ability to deliver the wages, cost-of-living increases,
ever-widening benefits, and safe working conditions that had been the
cornerstone of collective bargaining in the postwar years. Employers
fine-tuned the use of both legal and illegal tactics against workers and
unions in contract negotiations, in organizing drives, and on the shop
and office floor. The stability that workers had come to rely on seemed
to go into free fall, unchecked by government regulations or union
power. For many—even many in the labor movement—it was unclear
whether labor would survive this onslaught.

This book chronicles the twenty-month battle between the Steelworkers and the Ravenswood Aluminum Company (RAC). It is the
story of an international union already reeling from heavy losses in the
steel industry and desperately needing some solid ground. It is the story
of a tough and determined union membership, most of whom had spent
their entire working lives at the local aluminum plant.

Like so many other workers and unions over the last two decades, the
locked-out workers at Ravenswood were facing something unfath-
omable, even in the context of their toughest bargaining in the past.
Their plant, a flagship plant of Kaiser Aluminum when it first opened in
the 1950s, had brought a middle-class lifestyle to their families in rural
West Virginia. In the late 1980s it was sold in a leveraged buyout. The
workers watched as new management—driven to cut costs and increase
profits—slashed jobs, cut staffing levels, and let working conditions de-
teriorate dangerously.

In all of this they were up against a former plant manager, now CEO,
who had returned to the plant with a personal grudge against the union.
And after many months the Steelworkers would discover a tangled web
of overlapping owners and investors revealing that the company and
their destiny were ultimately controlled by a world-renowned financier
and metal trader, a fugitive from justice and notorious white-collar criminal hiding out in Switzerland. This would be no ordinary fight.

It would take almost two years, but in the end it was the Steelworkers
who held out "one day longer" than the company, defying the pattern
that emerged out of the 1980s of broken unions and divided communities. Without question, the victory was wrested from Ravenswood Aluminum by the tremendous determination and courage of the Steelworkers and their families. Even through the darkest times of the lockout, the
members of Local 5668 held on and held together, far beyond what rea-

son told them. In the twenty months of the lockout, only seventeen of their seventeen hundred members crossed the picket line.

But Ravenswood was not won by will alone. The victory was also the product of one of the most complex and sophisticated contract campaigns ever waged by the American labor movement. It was based on an immense amount of research, as well as careful thinking and strategizing about Ravenswood Aluminum Corporation, its owners, its customers, and its financiers. The Steelworkers never stopped thinking of innovative ways to put this information to use, constantly developing new strategies and tactics to pressure the company back to the bargaining table and get the workers and their union back into the plant. There were no bounds to these initiatives, not even national borders, as locked-out workers and their campaign traveled around the globe.

The Steelworkers did not make the mistake of partitioning off the research and the strategizing from the militancy and commitment of local union members. Virtually every action capitalized on the voices and solidarity of the members of Local 5668, which in turn inspired support from workers, unions, politicians, business leaders, and community groups across the United States and throughout the world.

When the Steelworkers returned to work on June 29, 1992, it was a tremendous victory for the workers and their families. Together they had saved their jobs and their union and in the process saved a community worth rebuilding. But Ravenswood was more than just a victory of aluminum workers in a small town in West Virginia. It demonstrated something that had been unclear for more than a decade—that labor could still win. Not just accidentally, not just against a weak employer—labor could win big against the largest of the corporate giants.

In so many ways the Ravenswood story represents the worst and the best of labor relations in the global economy. Here was yet another multi-national corporation wreaking havoc on workers and their community, all in the name of short-term profits to finance leveraged buyouts in industries thousands of miles away. But here also was a group of workers who found community, solidarity, and, ultimately, victory by reaching and touching hundreds of thousands of others in Eastern and Western Europe, Latin America, Canada, and across the United States.

Other unions won victories at the bargaining table and on the picket line in the 1980s and early 1990s. But it was the Ravenswood model—the research, the strategies, and the full participation of the members and the broader community, locally, nationally, and internationally—that represented both so much of what the labor movement had learned over the past two decades, and the possibility of what it could be in the years to come.

Ravenswood

1

HEAT STRESS,
HEAT STROKE

The heat came early to West Virginia in the summer of 1990. By mid-June temperatures had already been above 95 for days at a time and the air was heavy and thick with humidity. Inside the pot rooms at Ravenswood Aluminum, where ore is heated to 1,800 degrees Fahrenheit to make molten aluminum, temperatures were even higher—sometimes reaching 130 degrees. As Mike Bailes, a worker in the plant, would describe it later, mowing your lawn was hard enough that summer, but working in the pot rooms "was a hundred times harder."

In the past the company had always brought in summer relief to the pot rooms. But this summer was different. Under new management, workers were being forced to work double shifts—sixteen hours—sometimes five days in a row. For Mike Schmidt, working even one shift in the pot rooms that summer was unbearable. Working a double shift was even worse.

> You can't breathe. There's no air movement. It's still . . . and you're standing there and it's coming out of an 1,800–degree pot and it's put right at your feet. It's gassy, and you've got to stand right on top of it, and it's nasty . . . and they were forcing me to double, five out of seven days . . . and, you know, you come home and sleep four hours and didn't see the wife or the kids and then you go back to work. I mean, it was misery.

Friday, June 15, was one of the hottest days that month. Jimmy Rider was one of the fortunate ones in the plant; he worked in the air-conditioned cab of an overhead crane. But that Friday, like almost every other day that month, there were not enough workers to staff the pot-lines. No one had volunteered to work an extra shift. So, after he finished working the midnight shift in the crane, management forced Rider to work the next shift down on the pot room floor. It had been three years since he had last worked in the pot room and he wasn't used to the heat.

Two hours before the end of his first shift, Rider went to the medical department with an upset stomach and asked to be relieved of working the extra shift. The nurse refused and sent him back to work. As the morning dragged on, Rider repeatedly told his foreman, Swinesburg, that he wasn't feeling well and needed to go home. Each time the foreman ordered him back to work.

At ten A.M., two hours into his second shift, Rider came into the lunch room one last time and lay down on the floor. He told his fellow workers, "Boy, I'm too hot." Then Jimmy Rider, thirty-eight years old, died of a massive heart attack. His fellow workers and the company nurse tried to revive him, but none of them, not even the nurse, had been trained in CPR.

After the paramedics came to take Jimmy Rider away, work stopped. Company officials, who later claimed that Rider's death was the result of "an undetected non-work-related heart problem," seemed more disturbed by what they considered an illegal work stoppage than by his death. Rider's friends and fellow workers, still stunned from what had happened, were forced back to work.

Twenty-five-year-old Mike Schmidt, healthy and young but exhausted after working five double shifts in a row, was one of those ordered back into the unbearable heat. Just a few hours later, Mike himself collapsed on the floor, his brain swelling from heat stroke. The company nurse packed him with ice. Mike was afraid he was about to die. Forty-five minutes later his temperature was still above 102 degrees. Only then was Mike Schmidt permitted to go home. Later that day, a third worker collapsed from heat stroke.

In the days that followed, the union repeatedly met with the company to convince them to hire more pot room employees and eliminate mandatory overtime. Instead, the company threatened to shut down one of the potlines, laying off more than fifty workers, unless all pot room workers agreed to twelve-hour shifts.

With their members exhausted from repeated double shifts and still grieving over Rider's needless death, the union rejected twelve-hour shifts and told the company to shut the potline down. Restarting it would cost more than a million dollars. This was much more than it would take to quickly hire and train a summer relief crew, but still the company shut the potline down. Mike Bailes, one of Jimmy's friends, believes the company felt "they had to do that because, if they didn't, it would have been an admission of guilt that they had just killed the guy."

Three more workers died at Ravenswood during that long hot summer. The day after Rider's death, security guards Peter Baltic and Curtis McClain were overcome by carbon dioxide when they went to check out a fire alarm in a basement room. A month later, Dave Evans, an electrician with fifteen years' experience on his job and thirty in the plant, died of burns from an electrical box explosion.

Nothing like this had ever happened before at Ravenswood. In the entire thirty years the plant had been open, only a handful of deaths could be recalled. But under new ownership it looked like things were going to be different.

2

MACHINES IN THE GARDEN

Ravenswood sits on the eastern bank of the Ohio River in the far western part of West Virginia, a three-hour drive south of Pittsburgh. Here the rolling hills flatten out as they approach the river. The town is not much more than a single street tucked close along the river, crisscrossed by short side streets—imitating the railroad track that runs through town. It has a flat, almost Midwestern, feel. The main street running north out of Ravenswood is marked by the obligatory McDonald's, Pizza Hut, and a strip mall that contains the local supermarket, Foodland.

The land around Ravenswood was prime Native American hunting ground until claimed by Virginia colonists as their reward for fighting with the British in the French and Indian War. The most prominent among them was George Washington, who in 1770 laid claim to a large tract of land along the Ohio and Kanawha Rivers, which included the site that later became Ravenswood. According to Washington's diary, the land was prized for its "luxuriance of soil," plentiful wildlife, and water transportation up the Ohio and the Kanawha to the Monongahela River and frontiers beyond. Upwards of twenty thousand acres, this land was known for many years simply as "Washington's Woods."

South of Ravenswood, Route 2 follows the Ohio River out of town. A half dozen miles or so later, as the highway drifts away from the river momentarily, a smaller macadam road heads off toward the river. Here, where the flood plain broadens at a bend in the river, the Ravenswood aluminum plant sprawls along the bottomlands. Begun in 1954, it was one of a number of plants built by aluminum giant Henry J. Kaiser and the Kaiser Aluminum Company.

Kaiser was a colorful and successful industrialist who built an empire and a fortune during and after World War II. In the 1930s, his construction firm helped build the Hoover Dam and the Grand Coulee Dam, beginning what would be a lifelong pattern of government contracts. In the early 1940s, Kaiser ventured into the shipbuilding industry as a supplier for the war effort and by 1942 was employing more than eighty thousand workers in his shipyards. In 1943 he built a steel mill in Fontana, California, to combat what he felt was the monopolization of the industry by "Big Steel." Kaiser also operated Permanente Metals, a primary producer of magnesium during the war.

Encouraged by the U.S. government as part of an antitrust action against the aluminum giant ALCOA, the Aluminum Company of America, Henry Kaiser moved into the aluminum industry in 1946. As part of the agreement with the Justice Department, ALCOA was required to supply all the bauxite (the raw material for aluminum) that Kaiser needed. He began operations by leasing two government-owned aluminum facilities in Washington state and expanded rapidly through the 1950s. The Ravenswood plant was part of Kaiser Aluminum's strategy to compete head to head with ALCOA in eastern markets.

Like George Washington two centuries earlier, Kaiser chose Ravenswood because of its prime location near major waterways. Alumina (bauxite that has been refined) could be inexpensively hauled by barge from Baton Rouge, where it arrived from around the world. The abundant local coal supplies could generate electricity to power the plant, and three-quarters of the nation's aluminum market was within a five hundred–mile radius of the town.

Designed as an "integrated operation," the Ravenswood plant includes a reduction facility to smelt alumina into molten aluminum, as well as a fabrication plant to turn this molten metal primarily into "can stock" for beer and soda cans. The plant was one of many massive industrial facilities built in places such as Chicago and Detroit and in small towns across the South and Midwest in the 1940 and 1950s. The tall, majestic brick structures of the earlier New England mills were replaced by low, flat, nondescript buildings that maximized the flow of materials and personnel. Built before concepts such as outsourcing or just-in-time production had been dreamed up, the idea behind the Ravenswood facility was to transport barges of raw materials to the plant and then use thousands of workers and millions of kilowatts of electricity to transform the alumina into miles of aluminum to be hauled out by hundreds of trucks and railroad cars.

While the town of Ravenswood and the rest of Jackson County are by any contemporary definitions still quite rural, in 1954 local residents expressed concern that the new Kaiser plant would disturb their country ways. But Jackson County was not without its problems already. With no industrial base and little work, the population declined by 16 percent between 1940 and 1955, as sons and daughters of local residents looked for work elsewhere. Given its high level of unemployment, Jackson County had been designated as a "rural problem area" by the U. S. Secretary of Agriculture.

Conditions changed with the construction of the plant. Four thousand construction workers suddenly descended on the town. Trailer parks and apartment buildings sprang up as Ravenswood began integrating more than five thousand hourly employees, managers, technicians, and their families into a town of twelve hundred. They needed homes, utilities, police and fire support, schools, groceries, churches, and medical facilities.

Despite its pastoral, down-home lifestyle, some potential employees found Ravenswood wanting. "Due to the uncertainty of schools and hospitals and community services," wrote an engineer considering the move to Ravenswood, "we do not feel it is fair to our children to make a move of this kind."

Kaiser responded to the shortage of schools by building a $500,000 elementary school, leased to the town for $1 a year. A planning commission was established, the town's first zoning ordinance was implemented, and a number of developers began construction on new homes, with low-interest loans arranged by the Federal Housing Authority. Roads were paved and parks and recreation areas planned. By 1957 *Business Week* reported that "The difficulties aren't gone by a long shot—but now Ravenswood sees itself becoming a modern, well-planned city based around an industrial complex."

Not all the workers in the new plant lived in Ravenswood. Some settled in Ripley, the county seat; Parkersburg, the nearest city, forty miles north; and neighboring towns. Others commuted from Gallipolis and other Ohio communities across the river. Kaiser drew workers from all across the region. Dewey Taylor had played semi-pro baseball and worked in the mines. When he was laid off from the Chevy plant in Cleveland early in 1958, he came to work in the casting department at Kaiser. Bill Doyle had worked in the steel mills and had been to college before signing on at the plant in 1957. Dan Stidham, at age twenty-two, was one of the first hired at the plant in 1957. "When I went to work in the casting department, they didn't have concrete poured on the floors yet. They did have one furnace built and running, but the second one was under construction. The walls weren't completely up on it."

The workers at Ravenswood organized in early spring 1958 with the United Steelworkers of America. Many workers who hired on at Kaiser, like Bill Doyle, had grown up around the labor struggles in the coal fields, and the United Mine Workers of America (UMWA) played a big part in their lives. "My dad was a vice president of the United Mine Workers up in Pennsylvania. . . . I was born and raised in a union town, so I pretty well knew the ins and outs of the labor movement."

The local union chartered at Ravenswood quickly gained a reputation for toughness. "We are militant, very militant, but not radical, you know," former Local 5668 president Bud Chenoweth suggests. "What we wanted was to be treated fairly." Over the next years the union bargained for wages and benefits that brought a middle-class lifestyle to these workers in rural West Virginia.

The postwar economy had an almost insatiable appetite for the building blocks of an industrial machine that supplied expanding markets at home and abroad. Lightweight, rust-free aluminum was becoming extremely popular in both industrial and consumer products, and the industry grew and profited dramatically. This growth supported a level of wages and benefits that had been inconceivable in communities such as Ravenswood. "The Ohio Valley today stands as a glittering triumph for U.S. science and engineering," wrote *Reader's Digest* in 1963. "Out of the valley's gigantic new plants are pouring products that serve as the foundation for an ever more affluent America. In the years to come, the Ohio is expected to make even greater strides. But as the people in the valley take stock of all that has come this way so far, they can only regard it as a 20th-century miracle."

Conditions were not quite so idyllic inside the Ravenswood aluminum plant. The work was difficult and hot, especially during the long West Virginia summers. Tempers flared, grievances were filed, and employees were fired. Still, bolstered by improving wages, benefits, and working conditions, the overall relationship between Kaiser and Local 5668 was positive. The Ravenswood plant saw no strikes or major work stoppages in all the years it was run by Kaiser Aluminum.

Through the 1960s and 1970s, Ravenswood was not a community where class divisions were readily apparent. Labor and management lived side by side in neighborhoods that were coming of age in and around Ravenswood. They went to church together, sent their kids to the same schools, and all helped out with Little League.

Emmett Boyle, a young engineer who later would be promoted to reduction plant supervisor, lived on a street of a dozen modest 1960s houses built behind the union hall. Called "Nu Chance Drive," it was hardly an exclusive neighborhood. The house across the street belonged to Don Flanigan, an hourly employee at the plant, and his wife Marge. Marge recalls the Boyle family: "They were very friendly, very community-minded. I mean, they weren't the country club jet set. . . .

They had nice household furnishings but nothing above mine, as far as that's concerned. They borrowed our encyclopedias. He [Emmett] borrowed our lawn mower."

Boyle's youngest son Bobby became almost part of the Flanigan family. "Bobby had a lot of problems adjusting to moving here from Louisiana," recalls Marge. "He started coming over to our house. Our son was probably five years older than Bobby, maybe even older than that, but Steve and Bobby became friends. Steve was kind of his idol." As time went by, "it was not unusual for Bobby to come in after school and jump up on my countertop and say, 'Well, what are we having for dinner tonight, Mom?'"

But by the late 1970s this easy relationship between labor and management in Ravenswood and communities like it all across the country changed dramatically, when the world they had come to know started coming apart at the seams. As Barry Bluestone and Bennett Harrison wrote in *The Deindustrialization of America,* "Every newscast seemed to contain a story about a plant shutting down, another thousand jobs disappearing from a community, or the frustration of workers unable to find full-time jobs utilizing their skills and providing enough income to support their families. . . . The system that seemed so capable of providing a steadily growing standard of living during the turbulent 1960s had become totally incapable of providing people with a simple home mortgage, a stable job, or a secure future."

Americans sat by and watched as entire industries collapsed or were reduced to shadows of their former selves. Chrysler almost went under, the consumer electronics industry disappeared altogether, and the steel industry went into free fall. John Hoerr described the Mon Valley, the home of the steel industry in western Pennsylvania, in 1988: "The number of people who derived their income from steel had declined to less than four thousand, down from over thirty-five thousand in 1981, and from eighty thousand in the late 1940s. The mill towns, once so alive with the heavy throb of industry, now gave off the weak pulse of welfare and retirement communities. The degree of suffering caused by lost jobs, mortgage foreclosures, suicides, broken marriages, and alcoholism was beyond calculation."

Few of the tremendous postwar profits in the steel industry were invested back into the aging mills, as giants such as U.S. Steel—which changed its name in the mid-1980s to USX—turned instead to investments in energy and retail to diversify their portfolios. A new "bottom-line" ethos emerged which focused only on the next quarter's profits, rejecting long-term investment, research, and development. By the early

1980s, even the *Harvard Business Review* was concerned that we were "managing our way to economic decline."

In the face of a growing economic crisis made largely by its own hand, business turned on workers and unions. While workers had been junior partners in postwar success, they were now asked to become senior partners in failure. More and more employers turned to their hourly employees for concessions in wages, benefits, and longstanding work rules. The "compact" that marked postwar labor relations was gone. By the early 1980s it was clear that management had taken off the gloves.

Ronald Reagan fired striking air traffic controllers and through a series of appointments constituted a National Labor Relations Board (NLRB) that was explicitly anti-labor. And in a small mining town in Arizona corporate America pioneered a new model for operating during strikes. The miners at Phelps-Dodge, members of the United Steelworkers, had always bargained as a group with other unions and employers in "pattern bargaining" in the copper industry. As Phelps-Dodge pushed to abandon the pattern and roll back wages and benefits, miners in Morenci struck. Like most unions in the postwar years, they believed that a shutdown would demonstrate their seriousness and bring Phelps-Dodge back in line with the pattern agreement. But the company fought back.

Relying on the handbook *Operating during Strikes,* developed by Herbert Northrup, a former labor relations specialist from General Electric, Phelps-Dodge kept the facility open. Management hired massive numbers of replacement workers and a permanent security force. Aided by injunctions and court decisions, they waged an aggressive and bitter year-long battle against both the union and the community in Morenci. Once Governor Bruce Babbitt called in the National Guard, the strike was all but over. Josephine Rivas described the scene, "First the DPS [Department of Public Safety] cars came, one after another with their lights flashing. Then all the trucks. Then the tanks came through on platforms. It was like a war." The local union was decertified and the Steelworkers were beaten. While it seemed to be only a victory for management in one tiny Arizona community, Phelps-Dodge became the model for breaking unions in the 1980s, used dozens of times across the country over the next decade.

The ink was barely dry on the 1982 contract when Kaiser came to Local 5668 in Ravenswood asking for concessions. After expanding almost unchecked for three decades, the aluminum industry had been hit hard by the Reagan recession. Kaiser management demanded wage concessions and proposed to combine a number of positions in the pot room. Negotiations took place with the reduction plant supervisor, Emmett Boyle.

Boyle was convinced that the union contract had far too many job ti-
tles and restrictive work rules in the pot room. According to Steelworkers staff representative Dallas Ellswick, Boyle threatened to shut down
the last operating potline unless the union met his demands. "They
wanted to make some changes in the reduction plant, make some job
combinations," remembers Bud Chenoweth, the local president at the
time, "and they were telling me that was what the people wanted, you
know. And so I told him that wasn't what the people wanted. They kept
on, kept on. We've got to have these. We've got to have these." Boyle
kept needling Chenoweth, who finally challenged Boyle to hold a refer-
endum on the proposal. "There's one way to find out—we'll have a
vote." Boyle, inexperienced and hotheaded, agreed.

In a miscalculation that would cost Boyle dearly, workers voted overwhelmingly to reject his proposal. With his credibility undermined,
some months later Boyle was forced to leave. The way the union people
saw it, he was fired. He reportedly transferred to another job with Kaiser
in California. Boyle took it hard. Bud Rose, a committeeman at the time,
remembers, "I was one of the last ones to leave the room that evening.
[After the vote count,] he was sitting at the table by himself with big tears
running down his face." "They had purchased a nice home, and it was a
good neighborhood for them to raise their children," remembers Boyle's
neighbor Marge Flanigan. "And I think he thought his career would end
here in Ravenswood, you know, that he would retire from Kaiser."

Kaiser followed through with the threat and shut down the potline.
The local had to make concessions in 1984 to get the potlines running
again and the plant back in full operation. But in many ways the magic
of the Kaiser miracle was gone. Labor relations had been soured in some
fundamental way and Ravenswood would never be quite the same.

3

New Owners,
New Management

By the early 1980s the aluminum industry had stalled. The growth in demand for aluminum, which had hovered close to 10 percent per year in the 1960s, dropped to 3 percent. Martin Marietta, the sixth largest producer, got out of the business by 1984, and the rest of the industry felt the pinch of slackening demand.

These problems were partly a result of the recession that was devastating manufacturers nationwide. Aluminum was hit particularly hard because by the 1980s the metal had largely saturated its markets. The aluminum can industry grew more than 12 percent annually during the 1960s and 1970s, but by 1981, 80 percent of the potential market was already using aluminum cans. The same was true in building and transportation. As Charles W. Parry, president of ALCOA, suggested, "Penetration in some markets has gone as far as it's going to go." Yet the crisis in aluminum was not simply a product of the market.

In the 1980s American companies and major multinationals got caught up in a frenzy of buying and selling, gobbling up smaller companies and dismantling larger ones. As part of this massive corporate reshuffling, for example, U.S. Armco Steel was purchased by Kawasaki Steel, LTV and Republic Steel merged, and Japan's Nippon Kokan purchased a 50 percent interest in National Steel. This restructuring had nothing to do with a desire to strengthen American manufacturing. As corporate raiders from Michael Milken to Carl Icahn soon realized, there was a fortune to be made buying and selling companies. And, as Seymour Melman wrote in his 1983 *Profits without Production,* "Once corporate managers begin to treat part of their enterprises as money machines, the nature of the product becomes secondary."

The structure and control of the aluminum industry changed dramatically during the 1980s. *Aluminum: Profile of an Industry,* published in 1969, listed only eight domestic aluminum producers. These included industry giant ALCOA, followed in size by Reynolds and Kaiser. With the exception of Ormet, which was jointly owned by Olin Mathieson Chemical and

Revere Copper and Brass, all were solely-owned domestic companies, producing primarily aluminum. All this changed in the 1980s, as aluminum became a global industry with a whole new set of corporate players. Martin Marietta's facilities were sold to Australia's Comalco. Atlantic Richfield was bought by Canada's Alcan Aluminum. The aluminum industry was considerably more volatile and vulnerable to corporate buying and selling than the steel industry because aluminum, like copper and magnesium, is traded on the international metal markets. This means that individual companies have considerably less control of their fortunes than other manufacturers and are vulnerable to fluctuations and manipulations on the metals exchange. As one aluminum executive suggested, "You . . . have your fixed costs and someone who doesn't give a damn setting aluminum prices."

Henry J. Kaiser died in 1967 and what happened to the Kaiser Aluminum Company mirrored what happened in much of the industry. By 1986, dwindling markets and stagnant management found Kaiser $1.3 billion in debt. In this vacuum, an investment group led by Oklahoma businessman Joseph Frates III began a proxy fight to take over Kaiser Aluminum. Just two years earlier, in the kind of deal that defined the 1980s, Frates had taken over Kaiser's Steel Division. He turned around and sold the company, walking away with $14 million in fees and expenses.

Unlike the Steel Division, Kaiser Aluminum fought back, taking out full-page newspaper ads suggesting that Frates would dismantle and sell the aluminum business as he had steel. They even hired private investigators to look into Frates's private life. Despite these efforts, the industry failed to fall in line behind Kaiser and its chairman Cornell C. Maier. As one insider explains, "I would rather trust my money to Frates, who knows nothing about aluminum, than to the aluminum people who are wandering around the candy store."

While ultimately the Frates bid for Kaiser Aluminum proved unsuccessful, the company did not remain independent much longer. By 1988, two of the world's most influential financiers, Marc Rich and Charles Hurwitz, were fighting over control of the aluminum operation. Rich, a kingpin in the international metals market, dropped out at the last minute and Kaiser was purchased in a leveraged buyout by Charles Hurwitz and his holding company, Maxxam, Inc. Kaiser's aluminum operation was now under the control of one of the major corporate raiders on Wall Street.

After the deal, Maxxam had a $925 million note to pay off. As had been his pattern, in late 1988 Hurwitz decided to raise the cash by putting several Kaiser facilities on the block. These included a recycling center in

Indiana, a data center in Ohio, and the Ravenswood aluminum plant. In announcing the deal, 1988, Kaiser chair and CEO John S. Passman, Jr., claimed, "It is consistent with our strategy of selling selected assets, while at the same time preserving our competitive role as a fully integrated aluminum producer."

Yet by selling the Ravenswood plant, Kaiser cut its capacity to produce aluminum in half. Unlike in basic steel, where many of the plants "spun off" from major corporations were filled with antiquated equipment, the Ravenswood works was still a very viable operation. Although the plant was thirty years old, the process of making aluminum had not changed significantly in that time, and Kaiser had never delayed maintenance at the Ravenswood plant as was typical in many other manufacturing facilities.

The sale of Ravenswood may have represented a decision by Maxxam to retreat to Kaiser's traditional west coast base. But, as is the case in many leveraged buyouts, it is not clear that the decision to sell came either from an objective analysis of the aluminum industry and the Kaiser holdings, or Maxxam's interest in remaining a major player in the industry. Especially given the company's high level of indebtedness, it is more likely the decision was based on Maxxam's need to generate cash.

While not reflected in corporate statements about the sale, the decision to sell Ravenswood may also have been motivated by the militancy of the Ravenswood local. Beginning with Bud Chenoweth's confrontation with Emmett Boyle and through the struggles over concessions, the local earned a reputation for toughness. While Local 5668 bargained jointly with other locals for the Kaiser Master Contract, it was common knowledge that the local in Ravenswood meant business.

Dan Stidham was president of the union. Soft-spoken and stoic, Stidham does not fit the stereotype of a union leader. Unlike many local union presidents who tend to micromanage everything about the local, Stidham was a delegator. He picked good people and gave them space to move. Marge Flanigan describes Stidham as "quiet, very, very intelligent, I think. Quiet, methodical. He doesn't react to things and get angry. He thinks things through."

Stidham's alter ego in the local was the chair of the grievance committee, Charlie McDowell, an explosive, in-your-face leader. McDowell didn't care what people thought, and he spoke his mind. He loved the battle with management and reveled in the day-to-day struggle over grievances and work rules. "I enjoyed it. I loved it. I mean, I like competing against college graduate IR [Industrial Relations] reps. They were smart," says McDowell, who often beat them. "I did that time and time again."

In the *Local 5668 Newsletter,* Charlie's article on the proposed 1988 contract was "classic McDowell."

> It is now time for the Fat Lady to sing. Who is the Fat Lady? The Fat Lady is the membership who must tell the Company the same thing their representatives, Dan, Jack, Dewey and myself, have already told them. This contract does not meet the needs of hourly employees. The membership must be prepared to deal with those who try to threaten and intimidate you into voting for something short of what the membership deserves. . . .
>
> It is time for the Fat Lady (the membership) to stand up and say *NO* $1000.00 cannot buy my vote, because *I AM UNION AND DAMN PROUD OF IT.*

Joe Chapman, the USWA District 23 staff representative in charge of servicing the local, was one of those convinced that "Kaiser got rid of that place because of the local union attitude." When Chapman came to service the plant in 1977, the local was filing upwards of fifty grievances a day, resulting in "fifteen hundred-some grievances" for him to deal with at his step of the grievance procedure.

Local militancy had also erupted during ratification of the 1988 master agreement. Final votes on the master agreement were always pooled so that neither management nor the union ever knew how each local voted. But in 1988, faced with major concessions, the Kaiser locals insisted on knowing how their members voted. So, before sending the ballots to the international, each of the Steelworker locals counted the ballots at their respective union halls. Throughout the corporation the agreement passed by a 60 percent margin. But two locals voted it down. One was in Tacoma, Washington. The other was Local 5668.

The Ravenswood local was tough over local issues as well. In 1985, when Kaiser tried to set up employee involvement programs throughout the company, the Ravenswood local, particularly McDowell, wouldn't bite. "They sent in some really highly-paid consultants," explained Chapman, "and, basically, Charlie McDowell worked them over. He didn't believe it, and so he maybe did it a little deviously, but he opposed it, so he'd wine and dine them, and work at it, but he did not have any intentions of making it work."

The local was also strong on safety and health. Although safety committeeman Bill Doyle gives the air of an easygoing guy who would just as soon talk about fishing, which he loves, he dug in his heels about safety and health problems, which he had been meticulously documenting for years. If Kaiser had a good safety record at the Ravenswood plant, it was because health and safety reps such as Bill Doyle watched every step, kept track of every regulation and safety standard, and held management's feet to the fire. For John Molovich, USWA International Health and Safety Representative and former Deputy Commissioner of Labor in Indiana, the Local 5668 reps were "probably some of the most dedicated people that I know in this union as it relates to occupational safety and health."

While clearly some people both in management and in the international union believed that the "union ran the plant" in Ravenswood, and with their cockiness and militancy risked running it straight into the ground, others, like Molovich, saw the union's aggressive representation quite differently: "The company naturally thinks the local union is a thorn in their side, they don't want to cooperate. But there's a difference between cooperation and capitulation, and that's what the company wanted, and these [guys] wouldn't roll over and play dead. Ultimately they tagged them with the term that they were not cooperative. Well, thirty years ago, that was a good union person."

For all these reasons—the globalization trends in the industry, the need for quick cash, and the aggressiveness of the local union—the Ravenswood facility was sold February 7, 1989, to Stanwich Partners, Inc., based in Stamford, Connecticut. The new corporate identity, the Ravenswood Aluminum Company (RAC), was owned entirely by Stanwich president Charles Bradley, who specialized in these types of leveraged buyouts. Bradley bought the plant for $170 million, assuming approximately $180 million in pension liabilities. RAC also purchased the data center and the recycling center that Maxxam/Kaiser was unloading at the time.

Workers in Ravenswood were hopeful about the buyout. After the battle with Emmett Boyle over the pot room and several rounds of concessions, the workers looked forward to a change. "I, along with a lot of other people, kind of feel relieved that something happened," said Dan Stidham. But Stidham didn't expect major changes. "That's just my personal opinion, (but) if you're going to buy a company, you wouldn't necessarily want to come in and make a lot of changes and disrupt the workforce and give them reason to be alarmed."

Primarily a financier, Charles Bradley had no intention of owning and operating the Ravenswood Aluminum Company in the long run. Soon after the deal was made, Bradley transferred a significant portion of the company to Willy Strothotte, a Swiss metals trader. While Strothotte had considerably more experience than Bradley in the metals industry, he was clearly not the hands-on aluminum manager RAC would need if it was going to be successful. In June 1989, it was announced that Emmett Boyle would return to Ravenswood as CEO and as a major stockholder. An "aluminum man" all his life, Boyle would take charge of daily operations at RAC. Bradley and Strothotte had found their hands-on manager.

Most of the workers in Ravenswood had been sure that Boyle's career was washed up after the fiasco in the pot room. He was the last person they expected to take charge. It was also unclear how in such a short period of time he had gone from living on Nu Chance Drive to having the millions of dollars needed to control as much of the company as he did. No one knew how this sudden reversal of fortune had come about.

Marge Flanigan and her family had stayed in touch with Emmett and his family. Shortly after Boyle's reassignment to Kaiser in Oakland, Boyle's daughter Mary developed Hodgkins' disease. According to Marge, "They really had a terrible struggle out there. Emmett would call me on different occasions when he'd be back here, telling me what an ordeal the family was going through. He was telling me that Oakland wasn't what he had thought." Boyle missed the friends and community his family had in Ravenswood. Emmett told Marge, "Nobody cares. No one offered to help." At dinner parties nobody asked, "How's Mary?" "In Ravenswood," Flanigan says, "everybody circled around anyone that was ill . . . salaried, hourly—people didn't care." Marge also stayed in touch with Bobby. "We loved him like a son. When he left here, I would call Bobby and Bobby would call me and we couldn't talk because we were crying."

By April 1983, Boyle was back east as the president of Ormet, whose main reduction facility was located in Hannibal, Ohio, only a few hours from Ravenswood. The Boyle family relocated to New Martinsville, West Virginia, where they continued to stay in touch with the Flanigans. In 1986, Boyle, along with Charles Bradley, Willy Strothotte, and Thomas McGinty, formed Ohio River Associates and purchased Ormet from Consolidated Aluminum and Revere Copper and Brass in a leveraged buyout. By September Boyle's share of Ormet was valued at $12.6 million.

Although many of the workers thought Boyle's buy-in at Ravenswood came out of nowhere, he had already worked with Bradley and Strothotte at Ormet. USWA District Director Jim Bowen suspected that Boyle was involved in the RAC deal from the beginning. Bowen met with Charles Bradley to work out the close-out arrangements with Kaiser, including protecting the pensions of more than five hundred

Ravenswood workers. He asked Bradley point-blank, "Is Emmett Boyle going to be involved?" Bowen reports that Bradley answered, "Emmett Boyle's not involved." But he also added, "Not at this time."

Boyle and his partners did not buy into Ravenswood alone. Boyle, Strothotte, and Bradley also formed ORALCO Management, which controlled Ravenswood, Ormet, and Vialco, an alumina plant in the Virgin Islands. In a speech given at the West Virginia Business Summit on August 30, 1990, Boyle suggested that "Together, Ravenswood and Ormet can produce almost one billion pounds of primary aluminum at full capacity, making us the fourth largest smelting operation in the nation."

The sale of Ravenswood was not merely a paper transfer of ownership. Despite Stidham's belief that things wouldn't change, a whole new management team and an entirely different philosophy were brought to the plant. The first issues of the new company newspaper, *The Aluminator,* spelled out the new approach in the summer of 1989. Chief Operating Officer Don Worlledge wrote, "Ravenswood's future was in serious trouble before it was purchased from Kaiser. But RAC must be more efficient, more productive, and more concerned with quality if it is going to survive." Page one outlined cost-cutting measures and included a report from Boyle on rising costs. "This makes it essential that we bring down the costs we can control internally. This makes it essential that we work to make the quality of our products better than anything our competitors can offer. Now is the time for us to take the first steps toward restructuring our operations and strengthening our bottom line."

After he took over, Boyle quickly picked up where he had left off, implementing job combinations and staff cuts. The workers felt the change in ownership immediately. Many felt that Boyle had come back with a chip on his shoulder about what had happened in the pot room, and that his getting tough was in part retaliation for the way he had been treated. Before he left, Boyle had prophetically told Bill Hendricks, a worker in the pot room, "I'll be back. When I come back, I'll own this place." Hendricks remembers, "So then he bought the place, I like come clean out of my skin. And I told them guys, 'I want to tell you what he told me.' I said, 'This ain't funny, boys, take my word for it.' I said, 'I don't like the sound of this at all.'"

Jack Collins was the union committeeman (steward) in the pot room. "Kaiser was a company that had a little bit more compassion for human suffering . . . and RAC bought out and come in and had no compassion, no heart for human suffering at all. They could care less. And I think they came in with a vengeance to prove that, because of the fact that

Ravenswood always had a reputation. They always said the union runs the place. . . . The company had full control, but yet they wanted to make the people believe that we run it. And RAC come in there and they was going to break that union and they was going to break that power."

Boyle's changes were felt most acutely by the workers in the pot room. At best, smelting aluminum is hot, difficult, and dangerous work, and working in the pot room was the least desirable job in the plant. It was where all new-hires went, and everyone bid out as soon as they had enough seniority.

Aluminum is smelted in large electric furnaces. These "pots," as they are called, are steel shells lined with carbon. They are filled with alumina (refined bauxite) dissolved in a bath of molten cryolite. Carbon anodes are suspended in the pots and electrified, and molten aluminum moves to the bottom of the pot. The process generates tremendous heat, more than eighteen hundred degrees. Periodically, the molten aluminum is tapped from the pots and more alumina added from above. Pots are typically connected in a series, called a potline.

Regardless of an individual worker's job description, working around the smelting process is exhausting and hazardous. Mike Schmidt describes his job as "the hardest job in the plant, in that department. What you had to do was, you had to knock off carbon, which is sometimes as hard as concrete. And you had to do that on forty-eight of them. Then you had to shovel all of this into the pots. Then you had to go back through and fine clean them, and you're standing there and it's coming out of an eighteen hundred–degree pot and it's put right at your feet."

While management was combining jobs across the facility, in the pot room these cuts were devastating. As Jack Collins describes it, "When you combine jobs, that means somebody's got to go and the rest has got to take up the load. And when you're going to pick up that load with a twelve-pound sledge hammer and a thirty-five pound blue bar and eighteen hundred degrees looking you in the face, it's pretty rough. And through the job combinations it got rough and it just kept getting rougher."

From his perch in the cab of an overhead crane, Collins saw the impact RAC's changes had on the workers he represented.

I've seen people just lay down right there on that hot floor and say 'I can't make it.' They walk down to that work stop area and they can't walk back from here to that bridge to get into the cold air-conditioning because they didn't have the strength to make it. They just get so hot that their arms, their wrists, they just twist and pull. Their faces will turn to stone, like they're going into strokes. The same thing.

And in June, we had lots of problems in the pot rooms. It was a real hard summer. We had carbon problems coming out . . . and we had to burn off one right after another. That's the anode and the carbon separation. When that big black carbon don't come out with that anode, that means you've got to go in there by hand and get it. And we was having thousands of them. It just doubles up, triples up, and more, the work that you really have to do, and you're already doing a job that you can't hardly make.

Knowing that the job combinations, short staffing, and mandatory overtime were pushing the men in the pot room way beyond their limit, Collins organized against RAC's practices. That first fall after RAC took over, Collins began by organizing "informational picketing" in front of the administration building after workers finished their shifts. Fifty or more workers, mainly from the pot room, would show up carrying signs. Bill Hendricks remembers, "I looked up there in those windows and there were seven or eight videos going full tilt the whole time I was there. I knew right now we weren't dealing with your average company."

The picketing continued, but so did conditions in the pot room, including mandatory overtime. More grievances were filed, and in September 1989 Collins issued a letter to his fellow workers, listing three examples of company "abuse, threats, and retaliation." Collins went on to suggest, "I believe there is nothing the Company would like better than to crush this Union and strip you of the remaining working rights that still exist." He advised his members to press safety at the workplace, to work within their job description, to "work to rule" in a number of areas, and to get out of the company golf league.

One month later Collins received a written warning and a five-day suspension claiming that the letter "was designed, at a minimum, to incite employees to engage in unprotected work interruptions, disobedience of orders by supervisors, and the bypassing of the grievance procedure." After the company investigated his case, Jack Collins was discharged.

Collins was not an average union steward. He was extremely popular among the rank and file and most considered him in line to be the next local president. A ruggedly handsome man in his mid-forties, he had won his position as committeeman in the pot room in a plant-wide election. "They felt they could make an example out of me," says Collins. "Get me out of the plant for six or eight months, waiting on the arbitrator's decision. By that time they was going to have things pretty well settled. See, because they felt I was the leader, leader of the union, and if they could cut my head off, you know, it [the union] was going to die."

The loss of Collins was a tremendous blow to the union. In a local with leadership all over the age of fifty, he represented the hope and promise of the future. Charlie McDowell, shop chairman at the time, saw the dismissal as part of a larger strategy, "their first attempt to break the local union."

As Collins was escorted to the showers by management, he was followed by fifteen or twenty workers from the pot room. "Soon as I got out of the shower and went to my locker to get my stuff, here they come. They'd got showered too. They said, 'We're walking out.' And I said, 'No, you're not.' I said, 'You walk out of here the weeds will be growing up around the window sills of the union hall. I can tell you that right now, because they're going to do this.'" In his last action as a union leader in Local 5668, Collins, true to form, prevailed, avoiding what he and many others thought was a plan to force a wildcat strike.

After Collins was gone, conditions continued to worsen throughout the plant, especially in the pot room where Jimmy Rider died the following summer. Safety committeeman Bill Doyle could only stand by and watch as RAC dismantled the extensive safety and health program that he and Kaiser had begun in 1984. RAC managers no longer worked with the union's safety and health committee and conditions worsened so much that the management person responsible for safety and health left. "I was really friendly with the guy. . . . As a safety professional, he couldn't live with what they were attempting to do," recalls Bill Doyle. "He knew that, so he got out, and our procedure and policy just went further down the drain. We started fighting them every step of the way, back with OSHA, and fight, fight, fight."

RAC replaced their safety supervisor with Dave Caruthers, who had spent much of his career as a state highway patrol officer. Much to the chagrin of Doyle, Stidham, and McDowell, Caruthers spent more time on plant security than on safety and health. He seemed much more focused on the upcoming contract expiration than on the safety of the workers at Ravenswood.

The same summer that four workers died at Ravenswood, this new safety director spent his time making rather unusual security changes. According to Dan Stidham,

He put monitors on the top of the buildings, cameras, and installed a communications system in the guardhouse that we never had before. Now they can sit in there with the cameras and monitor everything that is taking place outside the gates or inside the plant. . . . They put doors up down in the plant that had never been locked before. Maybe they didn't need to be, but the company had been very lax and informal. If you wanted to go in and see the plant manager, you just walked in and saw him. Now today they got double-glass doors and double locks on them. Then on the outside of the building, they got it locked, so they got double security there.

But RAC management denied that these changes were part of any strike preparation plan. In Dewey Taylor's words, "They said that we're doing that, you know, to upgrade and replace windows for fuel efficiency, and the cameras was to protect our vehicles in the parking lot."

4

BARGAINING

As the local union negotiating committee began to prepare for bargaining a new contract with RAC late in the summer of 1990, one issue stood out from the rest. With the deaths in the plant and the deteriorating conditions in the pot room and across the facility, the safety and health of the workers at Ravenswood could not be ignored. The union wanted the labor/management health and safety committee restored. They wanted copies of injury reports and they wanted improvements in the way visits to medical were handled. Perhaps most important, in the aftermath of Jimmy Rider's death, the union wanted tighter restrictions on forced overtime, along with summer relief and increased staffing in the pot room.

Pensions were another major issue. The average age of the workforce was over fifty-two. Many workers had been at the plant since it opened more than thirty years before. So improving pension benefits was an overriding concern of a majority of workers. Other key issues included retaining current job classifications, restricting RAC's ability to contract out bargaining unit work, and holding on to the metal price bonus (an annual bonus tied to the market price of aluminum, established in 1983 in lieu of a wage increase). The union also had a strong desire to catch up with other aluminum plants in recouping wages and benefits lost in concession bargaining in the early 1980s. Kaiser workers had been particularly hard hit compared to workers at other major plants, and Local 5668 members wanted to get back on track.

Armed with these proposals, the union started negotiations with RAC in Washington, Pennsylvania, just outside Pittsburgh, on September 25 and 26, 1990. Chief negotiator for the union was Joe Chapman, the District 23 staff rep who had serviced the Ravenswood local since 1984. A West Virginia native, Chapman worked more than a decade in a Steelworkers shop in Huntington, was elected local president, and later went to work on the Steelworkers staff. Short and compact, with a trim beard, Chapman looks much younger than his early fifties. Despite his youthful appearance, in terms of servicing and negotiations, he says, "I'm kind of from the old school. My first staff guy I ever had, you didn't speak [in negotiations] unless he said so. He'd run the show." Joe felt the same way and was not the kind of union staffperson who worked to build a committee to handle negotiations themselves. He was in charge.

This was Chapman's style and the style of most USWA staff. In many unions, it is the local union and its officers who are signatories to the union contract and consequently the key players in negotiations. Under the USWA constitution, all contracts are with the international, not the local union. At Ravenswood Chapman and the international leadership, not Stidham or McDowell, ultimately signed off on all agreements with the company. This structure clashed at times with the desires of the local negotiating committee. Stidham was quiet, but not a pushover, and had strong ideas about what Local 5668 members deserved. McDowell saw himself as the key player in negotiations and his personality was not going to allow him to take a back seat

At the start of negotiations, Chapman was joined by District 23 Director Jim Bowen. District 23 included all of West Virginia, parts of Ohio and Kentucky, and a small part of Virginia. Bowen had worked at Olin Mathieson, which later became Consolidated Aluminum, before going on District 23 staff. In the 1970s, he had serviced the Ravenswood plant and later chaired the union's national Kaiser Committee "during the tough times in 85."

Bowen also played key roles in the Armco Steel and Wheeling-Pitt bargaining. But Bowen's forte is politics. A true politician in every sense of the word, Bowen operates with a canny combination of charm and bluster. One Steelworker staff member reported riding with Bowen in his car on the West Virginia turnpike at more than ninety miles an hour, late to a meeting in Charleston. When the staff rep warned Bowen that he might get caught speeding, Bowen replied that if the state police stopped him they would give him an escort straight to the Capitol.

Bowen had bargained with Boyle at Ormet in 1986. Chapman recalls, "In 1986 they had a strike out at Ormet. . . . Bowen had been involved. . . . And basically Emmett, after about four months, he rammed a contract right down their throat." Not one who liked getting beat, Bowen was out to get even. Some months after the Ormet contract was settled, Boyle was the scheduled speaker for a labor/management group in West Virginia. Bowen had bought several tables for Steelworkers members and, when Boyle got up to speak, twenty-five or thirty Steelworkers walked out behind Bowen, publicly humiliating Boyle. Joe Chapman recalled, "It was very pointed. Emmett didn't forget that." Marge Flanigan, recalling the incident, remembers Boyle as a very proud man: "He couldn't stand that ridicule."

The local union bargaining committee consisted of Dan Stidham, president; Charlie McDowell, grievance committee chair; Dewey Taylor, grievance committeeman in casting; Jack Wheeler, grievance committeeman in fabrication and maintenance; and Mike Bailes from the pot room. Wheeler, like Stidham, McDowell, and Taylor, was in his mid-fifties. He was reserved and rarely spoke. Bailes, still in his thirties, was the only young man on the bargaining committee, but even he had twenty years' experience in the plant. A lab technician in the reduction lab, Bailes first got involved with the union when elected to the 1980 bargaining committee.

This was the first time the Ravenswood local had ever bargained its own agreement independent of the master Kaiser negotiations. The committee spent all summer combining language from the master contract and the local agreement into one comprehensive agreement. Much of the work fell to Charlie McDowell. Coupled with his service on six negotiating committees since 1971, this work gave McDowell a strong sense of ownership about the contract and bargaining.

Earl Schick was the chief negotiator for RAC. A vice-president of ORALCO Management Services under Emmett Boyle, Schick was Boyle's right-hand man. Prior to joining ORALCO, he had had extensive dealings with the Steelworkers as a labor relations executive with USX. The management team also included Al Toothman, labor relations manager; Don Worlledge, RAC president; and several other ORALCO and RAC plant managers.

On the first day of bargaining, with the entire team present, Schick coldly laid out the company's concerns. Most important, he argued, RAC needed to hold down "Total Hourly Employment Costs" to survive. These included all wages, benefits, COLA (cost-of-living adjustments), and the metal price bonus. Schick also caught the local off guard by proposing to eliminate bumping rights between the reduction and fabrication plants. In the case of layoffs, very senior workers in one plant would be let go, while more junior workers in the other plant would keep their jobs. In effect, this proposal would lead to two separate contracts, seriously undermining the power of the union.

At this first meeting, the company threatened to keep operating if the union went on strike. They also promised (Jim Bowen called it a bribe) to consider starting up a third potline if a settlement was reached by October 31.

Schick divided the negotiators into an A team to address economic issues, which he would lead, and a B team led by Al Toothman that would
bargain over noneconomic issues. After lunch that first day, before the
union negotiators had even had a chance to present their proposals, they
got a sense of how difficult bargaining was going to be. Schick left. The
negotiations would be restricted to noneconomic issues with the B team
until Schick returned. He did not return until October 16, three weeks
later, when the company made its first economic proposals, which included freezing COLA and eliminating the metal price bonus.

The overall tone of negotiations was clearly being set by Earl Schick.
Paul Whitehead, the USWA attorney who joined the union team in the
last days of negotiations, described Schick's bargaining behavior as
"Prussian . . . very hierarchical, one person speaking." Joe Chapman
adds, "Well, Earl Schick and I can walk in here and you're about half-
smiling, that's the way he would be, and everything is friendly, and we
shake hands and everything. He goes around to that side of the table, sits
down to do business, and it's like, swish, I'm in my business mode now.
And he really gets deadly serious. There's no tomfoolery after that.
There's no smiles. That's just Earl."

It immediately became clear to everyone on the committee that this
would not be like negotiating with Kaiser. "I just felt threatened the first
day we went to the table," describes Dewey Taylor. "These people came
to the table and it was almost as if, here's our offer, take it or leave it . . .
we're going to run the plant with or without you. Right from the get-go,
that was their attitude. Just, this was the way it was going to be and this
was the real world." Unlike Kaiser, RAC was a privately held company,
where the management negotiators were bargaining for themselves, not
for the stockholders. As Dewey pointed out, every dollar they did not pay
the workers, and every dollar they took away from the workers, went into
their own pockets.

As the union committee members presented their proposals to management's B team, they passionately explained why health and safety was
their overriding concern. But Toothman categorically refused to bargain
over any of the pot room health and safety issues. "I'm not going to do
anything about the pot room until after November the 1st [after the contract expired]. . . We'll accept suggestions from you guys and cooperation, but we don't want to negotiate on that." Toothman said he was
waiting for the results of a company heat stress study, which would not
be completed until some time after November 1. Joe Chapman's response was that "there had been a death and workers were being hauled
out packed in ice and that was all the information the parties needed to

negotiate." But the company held firm and refused to bargain over the pot room.

RAC had not been idle in the meantime. While the negotiating committee, which consisted of most of the local's leaders, was away, RAC had been making strike preparations. Rank-and-file worker Toby Johnson explains, "We could see that they were getting ready for something 'cause they were clearing the brush along the hillsides, clearing all that out, and that was a big job. They worked on it a long time. And it's not a job you would do just to beautify the area."

The company also built a landing site for a helicopter. According to Joe Chapman, "They come to our contracting committee, and said, 'We're going to build a helicopter pad.' 'What for?' 'So we can get people injured in and out of here.' Well, we knew better than that." The helicopter pad was close to the administration building and nowhere near the most dangerous parts of the plant. So, while the bargaining committee was holed up in a hotel three hours north of Ravenswood, members of Local 5668 worked on finishing the helicopter pad, which they were pretty sure would be used against them if the contract was not settled.

One salaried worker recounted, "We saw the orders coming through for bulletproof vests, Mace, and Mace holders. Food and entertainment were ordered." These were no ordinary strike preparations.

As the contract expiration date came closer, RAC boarded up the windows in the main office building and installed a chain link fence around the perimeter of the property, which was no small undertaking. Heavy steel plates were installed in front of the massive electrical transformer station near the front of the property, surrounded by tractor-trailers filled with hay. Both the plates and the hay were to stop bullets from penetrating the transformers. The plant, especially the pot rooms, depended on the station for power. Even a relatively short interruption could cause the potline to freeze up, costing millions in repair and restart costs. With all these fortifications, the workers renamed the plant "Fort RAC." Al Toothman, RAC labor relations manager, was later to report that the company had spent more than two million dollars on strike preparation.

As the October 31 contract expiration approached, the union and company had made little headway. RAC still refused to bargain over health and safety, and made no economic offer until October 24, one week before the contract was due to expire. The offer was made at 2:05 P.M. and the bargaining session ended twenty minutes later. Over the next week, less than three hours was spent on economic bargaining. If the contract was going to be settled, bargaining would go down to the wire.

Throughout the day the contract was set to expire, the union and company exchanged several counterproposals, but precious little time was focused on economic issues. Dan Stidham had come down with bronchitis and lost his voice. They had taken him to the emergency room the night before but, in his typical stoic manner, he was hanging in until the contract was settled.

At 9:00 P.M. on October 31, the company distributed its final offer: elimination of the metal price bonus, freezing of the COLA, a 25-cent wage increase coupled with profit-sharing of 9 percent of pre-tax profits after $25 million, and slight improvements in the pension multiplier. The final offer didn't even mention the union's concerns on pot room health and safety.

Everyone on the bargaining team could see that RAC was not interested in reaching an agreement that night. The union bargainers returned after their caucus with a proposal they had used a number of times before, including in negotiations with Kaiser in 1988. They made an unconditional offer to work under the terms of the old agreement until negotiations produced a new contract, with a commitment to give RAC forty-eight hours' written notice of intent to strike.

Management left the room. With only half an hour left until the deadline, the local committee's spirits flagged. Their efforts over the last months added up to so little. They called down to the local union hall one more time to make sure everything was in order. As they had agreed, contract or not, everybody on the midnight shift would report for work.

Despite the company's elaborate strike preparations and intransigence in negotiating, both Bowen and Chapman fully expected Schick to accept the contract extension. Just the year before, they had worked out a similar extension with Schick at Ormet. The company had brought in busloads of scabs and security guards, but after an additional twenty days of negotiations, the parties reached an agreement without a lockout or strike.

As the time ticked away, Chapman became convinced that RAC was either going to accept the contract extension or force a strike. Although several committee members, especially Charlie McDowell, seemed eager for a fight, Chapman wanted to do everything he could to avoid a strike. If he had to, he wanted to force the company into a lockout. "I knew if I got them to lock us out, then under the laws of the state of West Virginia, we could draw unemployment. . . . I did not want to put a seventeen hundred-person place out on the bricks, believe me. I've had eight or ten strikes in my time, there's not been easy ones. That being a place that was not experienced, people was used to a good standard of living . . . they did not know what it would be like."

The offer to work under the current agreement was part of the union's strategy to force the company either to agree to the contract extension or to lock the workers out. In addition, Chapman had arranged with District 23 attorney Stan Hostler to instruct the night shift to report to work as usual, without picket signs and without any type of strike activity.

At six minutes before midnight, management returned. Schick rejected the union's extension offer and declared an impasse. Bowen leapt to his feet, protesting in his bellowing voice. He knew too well that an impasse was very different from management refusing to negotiate, and that the distinction as defined by law would make an enormous difference to the members of Local 5668. Under private sector labor law, impasse occurs when both sides have fully exhausted all avenues for good faith bargaining, reaching a deadlock on all issues. Once a true impasse is reached (as defined by the NLRB), management is free to implement its final offer.

Refusing to acknowledge Bowen, Schick stood up and left the room. The bargaining committee members were unclear what this meant. Was the company going to come back in at one minute before twelve with a counteroffer? Or was this it?

When the clock moved past midnight, clearly bargaining was over. No last minute phone call. No final offers. The negotiations were finished, at least for now. As they sat there for a few minutes to make sure that nothing else was going to happen, the committee was stunned.

As the news began to sink in, they realized that they had to get back to the local. They weren't on strike, but they also didn't have a contract, or even an agreement to work under the terms of the old one. They had no idea what would happen tomorrow in Ravenswood. So they tossed their clothes into bags and, saying good riddance to the Holiday Inn, got ready to hit the road to get back to Ravenswood. But when they called down to the union hall before leaving, they got bad news. The workers had been sent home. The union was locked out.

5

LOCKED OUT

As the clock ran out on bargaining in Pittsburgh, Local 5668 members gathered at the union hall back in Ravenswood, waiting for word from the negotiating committee. "We had a meeting about ten o'clock, and the mood was that the eleven o'clock shift wasn't going in," recalled vice-president Bill Doyle, who, along with Bud Chenoweth and the District's attorney, Stan Hostler, was in charge back at the local. Finally, Charlie McDowell called from Pittsburgh and told them that, whether the union settled or not, the midnight shift should report to work as planned.

The union was not on strike. But there was still no contract—not even an extension, "so it was a hell of a job out on the floor convincing those guys to go in," recalls Doyle. "Between Stan and myself and Bud Chenoweth and the guys that was here, we convinced them to go in."

Negotiations had gone down to the wire with Kaiser in the past, but at the last minute the contract had always been extended. In the end, they had always settled. Although RAC had acted very differently than Kaiser over the past months, as Bill Hendricks walked back into the plant that night, he and most of his fellow workers expected a settlement.

Tall and thin, Hendricks comes across Lincolnesque in both his gawkiness and his sincerity. He had hired on in 1973, twenty-two years old, "fresh off the street," feeling lucky to have found a good job where he "wouldn't have to worry about nothing the rest of [his] life." Thirty-nine years old, he was one of the youngest workers in the plant, and one of the youngest union activists. Hendricks had worked sixteen years in the pot room under Kaiser. Although most workers bid off the potlines at the first possible chance, Hendricks stayed. It was tough, hot work, but for Hendricks it was a "good place to fend for my family."

All that changed after RAC took over. The work got harder, the days seemed longer and, as a "safety man" for the union, he started worrying about the men. Then his good friend Jimmy Rider was killed on the job. Still, despite all that had happened over the past few years, Hendricks went in with the rest of the midnight shift thinking there would be a settlement.

Hendricks had been on the job for less than an hour when the workers were ordered to pack their gear and get out. "Everybody leaves, everybody," the foreman told Hendricks. "They [told] us we had to leave with no contract. . . . 'Communications broke down. You'll have to leave.' Just an hour before, they told us to stay." As Bill walked back to his locker, he caught a glimpse through an open window of several buses sitting in the salaried workers' parking lot, engines running. He suspected that the buses were filled with scabs. The buses, marked Lakeshore Tours out of Cleveland, Ohio, had been on the plant grounds since late that afternoon.

The workers took their time leaving the plant. In Bill Hendricks's words, they "kind of hem-hawed around." At one point workers blocked a doorway where some managers were congregated, "and they expected there was going to be a kind of, you know, fight right there in the hall." Finally, "everybody made their way to the bathroom, to the shower room. Everybody showered. And I stayed. I was the last one to leave the pot rooms. After everybody showered I checked every shower stall. I didn't want none of my boys standing on a commode seat. You know, some of them guys wanted to stay pretty bad, so I looked. And I was the last vehicle to leave the parking lot, me and Red Shoemaker."

The company made sure the workers left nothing behind. They had been handed a notice a few days earlier, telling them to clean out their lockers before they left on October 31. That night the foremen stood over them, making sure their lockers were emptied. As the last group of workers went into the parking lot, they were met by a wrecking truck waiting to tow away any workers' cars that had problems, or any cars left behind. It looked as though RAC was making a clean break with their current workforce, leaving nothing to chance. As the last union workers left the plant, the gate was locked behind them.

Not only were union workers locked out, the salaried and replacement workers were locked in. As instructed, salaried workers had reported at 7:00 P.M. that night carrying changes of clothes, toiletries, and a six months' supply of medication. At first the mood was almost festive. One of the salaried workers reported, "It was like a big party," because, like the union workers, most believed a strike "wasn't going to happen." Although the preparations hadn't been nearly as intense, in past negotiations with Kaiser they had also been told to bring in their clothes and medication. But no strike had ever happened.

This time, however, by 1:20 A.M., as the last union worker left the plant, managers walked through the offices telling everybody that "the union has not signed the contract so we're locking the gate." Some of the salaried workers from the data center walked into the parking lot and watched the union workers outside the fence. It was especially hard on the seven women whose husbands had just been locked out. As one describes, "You just couldn't believe that this was happening. It was like somebody in the family was dying. . . . And when I left, I said, 'I'll probably see you [her husband] at one o'clock.' Well, it was three days later that I saw him. . . . I mean you're still inside the fence but you could see. And it was really sad to see the guys sitting out, standing out there . . . because they were as dumbfounded as everybody else. I mean, it really was just sad. Because there they were, locked out."

The replacement workers started filing into the plant just as the last group of union workers was being escorted to the gate. Some had waited in buses through the long afternoon and evening and were growing impatient. Several encounters occurred between union workers and their replacements, through doorways and down hallways. But neither words nor blows were exchanged, since neither the union workers nor those brought in to replace them fully comprehended what was happening.

The salaried workers had already worked their full day shift before they came back in that evening. The company had set up a cafeteria and recreation room, but cots for sleeping were not set up until close to 4:00 A.M. Few got much sleep bunked on cots in unfamiliar surroundings. Blurry-eyed at 6:00 A.M., they couldn't miss the locked-out Steelworkers picketing outside the gate. The company security guards were everywhere—along the fence and on every roof. The salaried women in particular had been warned by the company not to go anywhere alone for fear of union violence. Yet, as they looked outside the fence, what they saw were their co-workers, their friends, and their neighbors.

After too little sleep, the salaried and replacement workers worked the first of many twelve-hour shifts. While the replacement workers were assigned to work in the plant, the secretaries, payroll clerks, bookkeepers, and accountants replaced locked-out maintenance and janitorial staff. The company rewarded them that first evening with a big dinner. Pool tables, basketball courts, and even televisions were provided, hooked up to two newly installed satellite dishes. In a display of generosity, management also provided a keg of beer, although everybody was limited to one glass.

After an exhausting drive from Pittsburgh on no sleep, the negotiating team arrived in Ravenswood at 5:00 A.M. They drove straight to the union hall, which is up on a ridge overlooking Route 2, midway between the town and the plant, just off Nu Chance drive, where Emmett Boyle once lived with his family. Fronted by a large rectangular parking lot, the union hall is a simple, one-story brick building from the 1960s, built originally as a community church. By the looks of it, it could have been an elementary school or a post office. It is not one of the union halls found in corporate office parks that are hard to distinguish from corporate offices. But it is also not one of the tired old union halls found in declining steel mill towns.

With its linoleum floor, the hall feels plain but spacious and functional. The furniture is pretty much mix-and-match—a few nice chairs in the president's office, older office furniture and homemade shelves here and there. This is not an office for show but for work—lots of filing cabinets, files stacked everywhere, a modern computer, and a fax machine whose high-pitched tone interrupts at random intervals. Most of the building is taken up by a large meeting room at the back of the hall. Like the rest of the hall, the meeting room is plain and functional. An old picture of the first president of the Steelworkers, Phil Murray, hangs on the wall, along with photographs of Steelworker district directors, including Paul Rusen, for whom the hall is named. Folding chairs are set up audience-style, facing a small low-riser and a podium.

When the committee pulled into the parking lot that morning, it was jammed with cars lining both sides of Route 2. The hall was packed. The mood was somber but focused. Prior to negotiations, the local had elected picket captains and mapped out the plant, but had not done much more planning in case of a strike. Now plans were quickly pulled together, picket duties assigned, and picket signs made. But, instead of "On Strike," the signs read "Locked Out." Many went right down to the picket line not having slept at all that night.

Some dressed in camouflage, as their union brothers in the UMWA had done in the long fight against the Pittston Coal Group a few years earlier. During the strike in southwestern Virginia in 1988, the miners, many of whom had served in Vietnam, adopted a uniform of camouflage. It not only reinforced the idea that they were fighting for their lives, but made identification of individuals by law enforcement difficult. Many of the Steelworkers at Ravenswood had supported the UMWA strike, and even those who were not directly involved were connected in spirit, given their longstanding connection to the UMWA and the struggles in the coal fields.

The picket line was organized in four-hour shifts, with the first workers reporting from midnight to 4:00 A.M. Not expecting a lockout, few had dressed for the picket line, and they shivered in the cold. Except for a handful of former UMWA members, most of those who picketed that night had never been on strike before. They really had no idea what to expect. Bill Hendricks explains, "It was dead cold. We didn't expect it to get that cold. And when we went back around the curb to go down there, I was freaked. They'd fired those portable generators. They had portable lights that lit up through the bottoms down there like it was New York City. I'm telling you, I've been driving to that plant at nighttime for years. On midnight shift, you couldn't see anything. . . . It was daylight around that plant. Beat anything I ever seen. . . . Matter of fact, you look in that light you could see fifty deer."

From the very beginning, the local union had no plans for massive picket line action. They planned to run the campaign by the book—no violence, no arrests, no civil disobedience. They would rely on negotiations and the courts to bring them justice and give them their jobs back. And they wanted to do everything possible so that this "labor dispute" was legally a lockout, not a strike.

This was not a mere issue of terminology. Locked-out workers cannot be "permanently replaced" and in many states, including West Virginia, they are eligible for unemployment benefits. In this way a lockout provides a union considerably more leverage. Consequently, it was in RAC's interest for the dispute to be identified as a "strike." In company press releases, and later in unemployment and NLRB hearings, RAC officials claimed that the union had never made a commitment to keep working under a contract extension. They knew the union was going to strike, they said, and had only sent the workers home to protect against the "chaos" that would have resulted as the union workers walked out "en masse" at the same time the replacement workforce was being brought in. They also argued that they were unable to agree to the contract extension, because the forty-eight hours' strike notice was not enough time to shut down a potline.

But the union leaders had never told the company they were on strike, not at the table when negotiations broke down in Pittsburgh, and not back in Ravenswood when they were sent out of the plant. Just a week before the contract expiration, the company had sent letters to all employees telling them "regardless of the status of negotiations, on November 1 there will be work for anyone who wishes to work." On October 31, union members had shown up for work and they were sent home. They were locked out.

A month before, as part of routine preparations for contract expiration, Dan Stidham had appointed former local president Bud Chenoweth as picket coordinator. That first night, as Bud huddled in the president's office with Charlie McDowell and Joe Chapman, they thought very carefully about how to set up their picket stations. As Charlie explains, "We did not want a strike sign anywhere in Jackson County."

The tensions increased, however, when the company brought in a forty-member paramilitary-style security force the night of the lockout. Equipped with riot shields, clubs, and tear gas, dressed in dark paramilitary-style clothing, they looked ominous. They also carried video cameras, constantly filming the activity on the picket line. The guards stayed at their posts, not joining other RAC employees except for meals, which they ate at separate tables. For those on the roof, buckets were passed up and down to bring them food.

Many of the workers saw the guards, who soon came to be known as the "goon guards," as a personal affront. The workers felt they had done nothing wrong. In fact, it was the company that had locked them out. They were only exercising their legal right to picket, and to be met with this kind of show of force was insulting.

Starting that first night, the goon guards seemed set on provoking a confrontation with the locked-out workers. They would challenge workers through the chain-link fence, calling them names. The union was equally determined not to play into the company's hand. To avoid the kind of sensational confrontation that the goon guards were clearly trying to incite, and the court injunction that would immediately follow, they pulled their pickets from the front of the plant on the second day of the lockout and moved them out to Route 2. As Joe Chapman tells it, the union had several reasons to move the pickets back. "When someone stands and throws an apple at you, pretty soon you're going to throw something back. So I didn't want that to happen. . . . I wanted to get the community, the public, everybody on our side. I wanted to put the company in the box of being the bad guy. And as quick as possible. And keep them there. And nail the lid shut on them."

Chapman and the local leaders also wanted to move the pickets back from the main gate to make them visible to people driving down the main road, not just to those taking the side road to the plant. As part of their effort to make their struggle more visible to the public, the locked-out workers immediately took their story to the media in the surrounding communities of Ravenswood, Parkersburg, Ripley, and Point Pleasant. By Thursday, November 2, the company had joined them in what one local paper called a "war of words and nerves." In their interviews, Joe Chapman and the local union leaders emphasized that they were not on strike and not at impasse—they were willing to keep bargaining and they were willing to keep working.

For its part, the company issued a series of press releases declaring that impasse had been reached and that given the "unrealistic contract extension proposal of the Steelworkers . . . the corporation believes that the only way to guarantee a continued and secure operation is to utilize the current salaried work force and to supplement these workers as necessary." The company also sent two mailings to the workers' homes, outlining their final offer and announcing that health benefits had been suspended as of November 2.

On Saturday, three days into the lockout, union members gathered by shift in three meetings at the Ravenswood High School gym. They came to vote on the committee's decision to reject management's final offer and agree to work under the old agreement. By a vote of eleven hundred to thirteen the rank-and-file stood solidly behind their committee.

Inside the plant four hundred salaried workers and nearly two hundred replacement workers worked round the clock to keep the plant running. For the first three days, they were locked inside behind boarded-up windows and doors. Their only contact with the outside was phone calls closely monitored by the guards. Over and over again the company kept reminding them of the threat of violence from the hourly workers. "It was like a prison. You couldn't see out. You didn't know what was going on outside. If it rained, you didn't know it. It's an awful feeling. . . . I thought, how long is this going to last? How long do we stay in this? And we didn't sleep. . . . I called him [her husband] a couple of times a day to see what was going on and to tell him I was okay. And we're still here and, no, I don't know when I'll be home. When they let us out, they let us out of prison."

At 7:00 P.M. on the third day, the salaried workers were told to go home for the night but to be back at 7:00 the next morning. Before the lockout they had worked Monday through Friday in day shift office jobs. Now they were put on twelve-hour shifts, 7:00 A.M. to 7:00 P.M., seven days a week, doing both their office work and the work of the locked-out janitors and maintenance workers, cleaning up the plant and cleaning up after the replacement workers. Nearly a month passed before any of the salaried workers were given a day off.

Down the road from the plant, Local 5668 members walked the picket lines. Although the guards and the replacement workers continued to try to provoke the picketers, the situation remained fairly calm. The worst moments came when the scabs would drive out of the plant waving their paychecks in the faces of the locked-out workers.

Negotiations did not resume until November 14, when both sides returned to Pittsburgh for two days of bargaining. This time they were joined by federal mediator Carmen Newell. But the union left Pittsburgh with a new company proposal that they felt offered even less than the one they had rejected two weeks before. The company sweetened their hourly pay increase from 25 to 35 cents an hour, but made no movement on health and safety and held firm on eliminating the metal price bonus and cost-of-living increases. The new proposal also backed down from the $1,000 lump sum signing bonus, which RAC now proposed not to pay until after 750 hours of work.

Although further sessions were scheduled with the mediator later in the month, RAC officials announced that, in light of the union's rejection of their latest proposal, they had no choice but to declare impasse and implement their final offer. Calling the dispute a strike rather than a lockout, RAC sent letters to all hourly employees inviting them to return to work on November 29 under the terms of the final offer that the union had just rejected. Meanwhile, the company expanded its efforts to recruit replacement workers through advertising in local newspapers. Even though every ad included the obligatory warning "Labor Dispute in Progress," by the end of November the number of replacement workers had climbed to more than four hundred.

The relative calm of the first three weeks of the lockout ended the last week in November. While hunting on her property on Thanksgiving Day, Karen Hughes, wife of a locked-out worker, climbed into a tree stand in a wooded portion of her property in hopes of sighting a buck. Instead, she saw one of the RAC security guards, dressed in camouflage, walking back and forth on the edge of the ridge overlooking the Local 5668 picket station near the north RAC entrance on Route 2. He took off running as soon as she yelled to ask him what he was doing on her property. As she climbed down the tree, she sighted another guard on the same ridge. Certain that they were RAC guards, she ran down the hill to get help from the union picketers. Although one of the guards got away before they could catch him, they found the other hiding in the weeds. He had a backpack containing food, a video camera with a telephoto lens, a two-way radio with codes for the RAC security guards, and an anti-riot tear gas grenade. He didn't resist their efforts to bring him back to the Hughes house, and within minutes the sheriff was called and the guard was arrested on trespassing charges.

Tension over the goon guards escalated on November 25, when they marched en masse to one of the picket stations. Dressed in full riot gear, marching behind Plexiglass shields, they demolished a picket shack, destroyed picket signs, and tore down an American flag that the Steelworkers had placed there. The company took this violent incident, along with other minor scuffles, and, turning it against the union, filed for an injunction with the Jackson County Circuit Court to limit picketing. On December 5, Judge Fred Fox agreed to limit the number of pickets to six per site, but not without issuing a stinging criticism of RAC's use of armed guards, whom he called "arrogant black-suited military strangers." Fox ordered the guards to stay on RAC property. He also ordered both sides to resume negotiations in his courtroom. A few sessions were held, but RAC refused to participate in any meaningful way. After three meetings that proved fruitless, Fox's bargaining order was nullified by U.S. District Judge Charles Haden, at RAC's request. He ruled that a "state judge is without authority to regulate conduct protected by the National Labor Relations Board." The federal judge let stand, however, the portions of the Fox order limiting the number of union pickets and restricting the guards to company property.

That same week, a decision from a three-judge panel for the West Virginia Department of Employment Security Board of Review dealt the locked-out workers a devastating and unexpected blow. The administrative law judges denied unemployment compensation to the locked-out workers. They ruled that "the evidence presented does not support the establishment of a lockout by RAC" and argued that the union had been engaged in a strike since November 1. Despite evidence that the parties were not at impasse and that union workers had reported to the plant prepared to work on the night of October 31, only to be sent home, RAC had prevailed.

The decision was a huge victory for the company. Inside the plant, management was jubilant. One salaried worker reported, "When they first announced that [the locked-out workers] weren't going to get their unemployment . . . they just rejoiced. 'They don't deserve it. They walked out. They walked out. They don't deserve anything. Nothing.'"

Not only did the locked-out workers now face weeks and possibly months of economic hardship, but the judges' decision paved the way for RAC to fire the "striking" union members and in their stead hire permanent replacements. Within days of the Review Board decision, the company issued a press release announcing that it would permanently retain four hundred workers hired to replace striking union members and "if members choose not to return to work, the corporation will be looking for several hundred more new employees."

In the aftermath of the Review Board decision, the union went back to the bargaining table prepared to make significant concessions, in hopes of jump-starting negotiations. The union sent a full committee with "clothes packed to stay and negotiate," but were disappointed to be met only by Don Worlledge and Earl Schick from RAC. Still, the union went ahead and presented a dramatically revised proposal, which included giving up the metal price bonus in exchange for a restoration of

1983 wage rates, dropping the demand for yearly wage increases, and dropping half the remaining noneconomic proposals.

But RAC took fewer than fifteen minutes to consider this sweeping new proposal. Schick rejected it out of hand and cut off negotiations without further discussion. Clearly, RAC was not interested in a settlement. In fact, the company was doing everything possible to break the union and ensure that it could replace its unionized employees with a nonunion work force.

6

HOLDING TOGETHER

In the first weeks of the lockout, some local leaders were confident, almost cocky, that the whole thing would be resolved within a few months. After all, they were a highly skilled and experienced work force. True to form, no one was more sure than Charlie McDowell. "I never felt for a minute they'd beat us. You cannot take the greenhorns over there and produce metal out of the plant that you can sell. You just can't do it."

Even though Local 5668 had never been on strike before, union leaders believed that if they put up their picket lines, held together, and told their story to the courts, justice would prevail and the lockout would be over quickly. However, by the middle of December, with picketing severely limited, no talks in sight, and more than six hundred scabs working in the plant, it was starting to sink in that this might be a long fight.

Things came to a head on a chilly December morning when former local president Gene Richards organized an unauthorized union meeting in Washington Lands Park. A few weeks before, Richards had stood up at a meeting at the Local 5668 union hall, calling on the union to return to work and continue to negotiate. That night he was shouted down by Stidham and the rest of the crowd. But on December 5, 1990, he tried again. This time Richards hoped to convince a substantial group to cross the picket line and go back to work. More than a hundred Ravenswood workers gathered in the park to hear what Richards had to say. Not long after the meeting started, word got back to the hall and a group led by picket captain Bud Chenoweth, a former local president himself, went down to see what the ruckus was about.

One of those who came down from the hall was Gene Lee, a veteran in the plant who had been with the union since the beginning. As Lee saw it, things could have gone either way that morning.

There was a pickup truck backed in there and Gene Richards was in the truck bed and I thought, uh-oh. . . . However, soon afterward Bud Chenoweth joined in. Before you know, Bud's in charge of the rally. But Mr. Richards wasn't going to be pushed totally aside. He eventually got a petition, wanting people to put signatures on it, present it to the local about doing something and getting us back to work. And so, after he said a few words about that, he was asking people to sign it. I said, 'You sign it, Gene. You wrote it.' I heard someone else over there cussing, 'You sign it, Gene, that'd be enough.' And the thing begin to wheee, go like that, a lead balloon.

Just then a company helicopter swept out from the plant and circled three times over the park, "to count the cars," according to Gene Lee. Later the union learned that Richards had promised the company he would be able to convince 250 workers to join him in crossing the picket line, and they had sent the helicopter to check on how he was doing. They saw the crowd, but what they couldn't see from the helicopter was that Richards' "unauthorized meeting" had turned into a rally of solidarity for the local union leadership, culminating in a drive-by past the RAC plant gates.

Two months into the lockout, all but thirteen of the seventeen hundred workers stood firm with the union. Yet, as Dan Stidham and the other local leaders knew, the union was going to have to do something more if it expected people to stay out, especially once their bills started piling up. To win, Joe Chapman told them, "We've got to keep food in their belly. We've got to keep a roof over their heads. We got to keep the house warm."

The local leaders had planned on unemployment benefits to take care of basic needs. But in the dark days of early December 1990, they recognized that months might pass before unemployment checks would come. Health insurance was just as important. The second day of the lockout, RAC informed the workers that their benefits would be cut off if they did not agree to return under the terms of the company's final offer. Workers could continue their coverage by paying hundreds of dollars each month, far more than most locked-out workers could afford. Pressure had been building for the union to do something about health insurance.

For James Watts, a forty-year-old union member stricken with cancer and facing medical bills in the tens of thousands of dollars, fear of losing his medical coverage was too much. Despite coming from a strong union family, and even remembering that the union had pitched in to help him and his family when he first came down with cancer the year before, Watts crossed the picket line on December 18 and returned to work. Others with serious health problems stood firm, but union officers knew they needed to do something fast to hold the members together.

With this in mind, Dan Stidham appointed financial secretary Bud
Rose to coordinate the establishment of an "Assistance Center." Rose
had come to work at Ravenswood in 1960, just two years after the union
came in. For his first fourteen years in the plant, Rose had no involvement in the union. But in 1974 he joined the safety committee and
within six months moved up to safety committee chair. Three years after
that, he became the local treasurer. At the time of the lockout, Rose was
the financial secretary, as well as a committeeman in the cold rolling department. A soft-spoken, congenial man, he was better suited to union
administration than to fighting with management on the shop floor. In
many ways he rounded out the team of Dan Stidham and Charlie Mc-
Dowell. He brought to the Assistance Center his organizational skills and
fiscal experience, honed first as treasurer and then as financial secretary,
as well as his generosity and compassion. As Stidham later recalled, Bud
Rose was a "people person, well suited for the tough assignment of equitably and sensitively providing the very proud and self-reliant Local 5668
members with financial assistance and moral support."

Rather than use the international strike fund to send individual checks
to each worker, as unions typically do during strikes and lockouts, the local leadership decided to pool their resources and base decisions on need.
Stidham describes it this way:

> The International Strike Defense Fund money came in and off the top
> we went out and bought food. And then we bought insurance for every-
> body, a basic minimum policy. Each member was getting one hundred
> dollars [a week from the International]. After you took off the food and
> insurance it was something like $76, $77 left, which all went in a pool.
> Each member didn't get the allotted amount, it was based on need. Some
> members' needs were greater than others'. There were probably members
> that never took a dime from up there. Some got everything they could
> get. Some maybe lived better than they did when they was working, or as
> good. . . . So it let that pool build and we were able to help some that
> were in more dire straits than others.

The fact that approximately a third of the locked-out workers had
Kaiser pensions to fall back on helped the local more equitably distribute
strike benefits. Starting in 1989 when Maxxam/Kaiser sold the plant,
nearly six hundred of the more senior employees still working in the
plant began to receive their monthly Kaiser pension checks on top of
their hourly rate. At the time of the sale, this had created deep divisions
within the ranks, captured in the angry graffiti on restroom walls about

"double-dipping." Now this pension money helped cushion the impact of the lockout for many, and allowed the scarce strike benefits to be concentrated on those in the greatest need.

Initially, the Assistance Center operated out of the back of the union hall. But with all the activity in the hall, there was no way to give the members the privacy they deserved to discuss their financial needs. "We did some checking around, and the shopping plaza up here had several empty buildings in it," Rose recalls. "We contacted the people . . . and they were very accommodating. They rented us that whole office up there for $286 a month. Eight or nine hundred dollars was the going rate, but he was very sympathetic to the cause, and he let us have it. Of course, we promised to do a lot of work in between time."

Neither Rose nor Local 5668 treasurer Glen Varney, whom Rose had asked to become co-chair, had any experience running this kind of operation. But they had good instincts as organizers and surrounded themselves with a core of a dozen or more members who would become stalwarts. Chapman had given them a book outlining legal issues in distributing strike funds and other assistance, but it was up to Rose and Varney to put together guidelines and a system for determining how the money and other donations would be distributed.

They also had to devise an office procedure to deal with the hundreds of workers in need of assistance. Learning by the seat of their pants, they assigned workers a specific day of the week to report to the center. They even devised an "intake system," where workers would sign up with a receptionist and then be referred to one of four interviewers who gathered all their relevant information on index cards.

As the holidays approached with no unemployment benefits in sight, more and more people came to the Assistance Center. The union began with a lump sum of $300,000 in strike benefits from the international union. Cash and food donations soon began to pour in from other area unions, including twelve hundred loaves of bread from members of the Bakery, Confectionery and Tobacco Workers union in Parkersburg. But Rose and Varney knew that, with seventeen hundred members, those initial funds would only last so long. They would need to build up a supply of money and food for the long haul.

Five days before Christmas, hundreds of Local 5668 members lined up in the Ravenswood Plaza parking lot for free hams and turkeys from the Assistance Center. As they waited, a cheer began to ripple through the crowd. They had just received word that the Board of Review had voted to overturn the previous unemployment ruling, on the grounds that the dispute was indeed a lockout, not a strike. RAC's last-minute appeal for an injunction was rejected. The locked-out members would now be eligible for twenty-six weeks of unemployment benefits retroactive to November 1. The holidays had come early for the members of Local 5668.

For several days the staff from the unemployment office set up shop in the Assistance Center and, according to Rose, workers were lined up "clean down over the hill" waiting to fill out their unemployment papers. Shortly after that, the local's executive board decided to provide Blue Cross/Blue Shield major medical insurance for those who needed it. Once again, for several days members waited in long lines to talk to the Assistance Center volunteers. By that time, several hundred people were being processed through the Center each day. Rose had no idea when the lockout might end, but he knew that unemployment would last only twenty-six weeks, with perhaps a ten-week extension. So the need to fund, supply, and staff the Assistance Center would continue.

The Center had started rather modestly, assisting with house, car, and utility payments, and providing a small food certificate. But even these decisions were hard. As Varney describes the dilemma, "You have to say, 'Did I do right here? Or should I do that? Did I make a car payment last month? Should I make this over here?' So all these things go through your mind, and people don't realize it." Rose explains, "It's really heartbreaking. This one particular day I remember real clear. Periece Morgan come in and he sit down. He said, 'I just about lost it out there. . . . This guy come in and had this stuff, and I told him we'd help him take care of it. He sat there and started crying. I had to leave, I couldn't take it.'"

Major bills were being taken care of and no one was losing his house or car. Like the union hall, the Assistance Center became a place where people gathered to catch up on the latest news from negotiations, to get an update on what Emmett Boyle and the scabs had been up to, or just to socialize and have a few laughs. While activities up at the union hall may have kept members informed, the Assistance Center allayed their fears and comforted their families.

The Assistance Center was also where the small group of women members of Local 5668 focused their activities. Of the seventeen hundred locked-out workers, only thirty-two were women. Most were skilled trades workers whom RAC had hired in the late 1970s as an outgrowth of a federal lawsuit for women to receive on-the-job-training (OJT) as maintenance mechanics, pipefitters, and electricians. In fact, almost all the women in the plant had started in the 1970s.

A small, feisty woman, millwright Janice Crawford looks much younger than the mother of grown children that she is. Part of the third OJT class of women, Crawford felt she had had a much easier time than the women who came before her, who faced more hostility from their male co-workers. Having spent years helping her husband in the body shop in their garage, Crawford had worked with tools before entering RAC, which made the transition easier. The only woman on her shift, she got along well with the men on the crew in part because, as she says, "I was real small, and I could get back in the greasiest, dirtiest, tiny little holes to get in places the men couldn't get into or get their hands into."

With the exception of Judy Cowan, who had been the local EEO (Equal Employment Opportunity) officer for many years and the grievance chair in the technical unit, none of the women had been very active in the local. The lockout hit them especially hard. At least a third had children at home, many were single parents, and others were married to men who worked in the plant, so that both wage earners were locked out. Early in the lockout the women became actively involved. For many, such as Crawford, involvement started with picket duty and moved on from there:

> Of course I started with picket duty. That's how it all started. And then we got to talking about the Assistance Center and I said, 'Just give me a call if you need anything.' Well, she [Janey Collins] called me one day and she said, 'We're going to take a run to the Food Bank.' And I said, 'Sure, let's go.' So that started the ball rolling. We went to the Food Bank the next day and learned how to do that. We took classes on that. Then we started setting it up and it just, it snowballed. It just got more and more. It started out with two days a week and then I went to three days a week. Then I went to five days a week. Then I moved up front to interview.

Even with the Assistance Center, that first winter was a difficult time for the locked-out workers. Dan Stidham describes the mood this way:

> Well, you've got a mixture. Bitterness. Bitterness toward the company that they'd do something like that to their long-term employees. You got bitterness and hatred against the scabs that come in and steal your job. They're going in and working . . . drawing a paycheck while you're not. They'd come out and wave their paychecks at the picket station or when they'd go by the union hall.
> People didn't have money to buy groceries. You'd go in a grocery store and see one of them cashing a RAC check; it just went all through you. I don't know, there's just so many different things that you'd encounter like that. It causes hostility and bitterness. Relatives of some of our peo-

ple locked out went in and took jobs. We had a member or two here that their sons went in and went to work. We had one working, took his dad's job, and his dad's out here on the picket line. That was a hell of a thing. We had brother against brother. We had two brothers working in there. One come out and stayed out, and the other went in and scabbed.

The workers' wives were no less confused and troubled by the lockout than their husbands. Marge Flanigan remembers:

Well, I'll tell you, in December we had one gal that was ready to go over the edge. And she came into the union hall and she begged for us to do something. And what she wanted to do was to go to the plant and have a prayer vigil. . . . We parked our cars at the airport. . . . We had candles and video cameras and we had a prayer vigil and we sang some songs, being filmed by the guards. And when we left we were right across from the plant . . . and they had left the plant gates open. And we walked . . . towards the plant . . . to make an impression that we just weren't afraid of them. So we walked across the road. They slammed the gates shut and within three minutes we had state police on us. They said that we had stormed the plant. Our candles turned into dynamite and that we were going to take over the plant. There was like twenty-five, thirty women.

And with this spontaneous action, the wives of the locked-out workers became involved. Realizing that the holiday was going to be difficult for everyone, several women members, together with some of the wives, organized a Christmas party in early December, with toys for the 740 children of the locked-out workers. Judy Cowan, Janice Crawford, and others asked unions in the area to set up gate collections to raise money to buy toys. The group also solicited donations for the party from more than a hundred area churches and businesses. Marge Flanigan, Emmett Boyle's former neighbor, got involved. "I had gotten rid of all my furniture before the lockout and the basement was totally empty. As donations started coming in . . . the basement was stacked to the ceiling. So I got real involved, accidentally, by living real close." The Christmas party was a huge success, one of the first social events since the lockout started.

Buoyed by the success of the Christmas party, the workers' wives decided to organize more formally into the "Women's Support Group." Marge Flanigan and Phyllis Fizer, wife of locked-out worker Joe Fizer, began calling wives, urging them to come to a meeting. "We met the first time, officially, probably in December of 1990 at the union hall," explained

Flanigan. "And it's just a group of women that got together and realized that families are falling apart. They're not going to understand labor disputes, we have lost friends, the families were pitted, our neighbors, the total devastation, the same thing as . . . if a tornado destroyed this community. We could rebuild it together."

The hall was packed that night with more than three hundred women. Clearly Marge was not the only spouse who felt total devastation. Many women felt isolated and uninformed, most knowing little about the lockout or unions. The Women's Support Group helped fill that void and gave the women a space to share their ideas, their problems, and their fears. The group also helped bolster the husbands, as they too were learning about solidarity. Support Group member Sue Groves suggests, "[It] . . . gave us something to grab hold of. At times we felt that, you know, what can we do? We were beginning to realize that this was going to be a long thing, and that some husbands don't talk to their wives, they don't share things. When this started, it was a source of information, a source of comfort. We shared and cried together, and we relaxed together, and it gave us something to work for."

But, as in the Flint, Michigan, sitdown strikes in 1937, and in the recent strike by miners against the Pittston Coal Group, the women were much more than a support group. In Flint the workers' wives, the Women's Emergency Brigade, in addition to running the soup kitchens to feed their husbands sitting in the plant, marched "with babies and banners" outside the shut-down GM plants in a direct challenge to both company goons and government troops.

At Pittston, the women's support group named themselves the "Daughters of Mother Jones." Mary Harris Jones, known during her lifetime as Mother Jones, was one of the best-known radical labor organizers of the late nineteenth and early twentieth centuries. She was involved in the 1880s with the Knights of Labor and was one of the founders of the Industrial Workers of the World (IWW) in 1905. She was best known for her strike support work, particularly with miners in the southern Appalachia coal fields.

Like their sisters in the Daughters of Mother Jones, the women of Ravenswood knew from the beginning that they needed to do more than prepare food and run Christmas parties. For Phyllis Fizer, one of the original organizers, their work was important because "we needed positive thinking and positive actions. We had to do everything we could to support these men and women locked out, and most of all, control our emotions and put them to good use."

One of the Support Group's first actions was a "Thank-you Rally" in early January to acknowledge area businesses, unions, and individuals who had donated goods and services that first winter. Close to a thousand Local 5668 members, wives, and supporters gathered on a snowy Saturday in the Ravenswood Plaza parking lot. Marge Flanigan explained, "We wanted to let our husbands know that we're standing behind them. We're in this together. So many bad things have happened these last few months, we've decided to try to turn some of the negatives into positives."

Something did feel different about that rally. When the women got involved, they moved the struggle beyond the locked-out workers to their families and to their community. As Dan Stidham told the crowd, "I think the workers and their families found out today that we're closer now than we were thirty years ago. The togetherness and feelings of friendship and solidarity are the good things that have come out of all this. . . . At least when we go up there [to Pittsburgh for bargaining], the company will know that we've got the support of our workers, our spouses, and probably 75 percent of the community."

This spirit was also evident at the "Rally in the Valley," when more than six thousand union members and supporters gathered at the Charleston, West Virginia, Civic Center on Sunday, December 30. More than twenty-five hundred members of Local 5668 families left the union hall to travel by car caravan to Charleston. Most were dressed in bright blue sweatshirts which read "Fort RAC, Union Solidarity" on the front and "Go Steelworkers" on the back. On the road to Charleston they were joined by thousands of other union members from Ohio, West Virginia, and Kentucky who came by the busload and carload through a driving rainstorm.

At the rally, with Jim Bowen as master of ceremonies, the locked-out workers told the story of court battles, the scabs, harassment by goon guards, and winning the battle for unemployment compensation, and listened to the stories of other unions who had won hard-fought victories after long strikes. They heard remarks from USWA international president Lynn Williams and Cecil Roberts, vice-president of the United Mine Workers, who exhorted the crowd, "Before unions came along there were two classes of people in America, the rich and the poor. Unless union members pull together and fight, it's going to be the rich and the poor again. And you ain't going to be rich."

The workers and their supporters returned to Ravenswood that night buoyed by their success and by the stories they had heard from the other unions. Rank-and-filer Larry Milhoan told a *Jackson Herald* reporter, "The support we received today makes me feel a whole lot better. We know that we are no longer just eighteen hundred workers against the world. There are others who have been through what we are going through and won."

7

CIVIL WAR

As fall turned into winter and the lockout dragged on, Emmett Boyle and the scabs he had hired became the focal point of the workers' anger and preoccupation. There was no middle ground. Either you supported Emmett Boyle and the scabs, or you supported the locked-out workers. The same winter that the Gulf War erupted in Kuwait and Iraq, their peaceful valley too had turned into a battlefield. Bill Hendricks declared, "It's like a little old civil war right here at Ravenswood. They talked about Kuwait. Well, this is little Kuwait."

In Ohio and West Virginia towns such as Sardis, New Martinsville, New Matamorus, and Powhattan Point, signs sprouted in store windows: "We support the locked-out workers" or "No scabs allowed." Businesses without those signs, particularly those frequented by RAC supervisors and replacement workers, were boycotted by union members and supporters. Boyle, in turn, sent letters to all the businesses who publicly supported the locked-out workers, threatening that RAC would not forget their actions after the "labor dispute" was resolved.

Despite pressure from his sons to continue to do business with RAC, Frank Young, the owner of Red Barn Wreckers, refused to service the company or the replacement workers. "My feeling is that the issue of justice is more important than the issue of money with me. On my part, I'm not going to use my trucks, and my business, in support of that scab operation. What is happening here is wrong. It should not be happening. People who have worked there twenty-five and thirty years, and in some cases are only months away from retirement, are being locked out from their futures, from their future work, and from their future retirements."

The tension between the locked-out workers, "locked-in" salaried workers, and the scabs touched every aspect of daily life. For the union members, who had spent their entire adult lives working in the plant, the company and the scabs had stolen not only their jobs but their community. "Well, you see a strange face or a new license plate in town, you think right away, 'there goes a damn scab,'" explained Local 5668 member Leonard Brown. "So you don't talk to people, you don't make new friends. People quit going to church because of the conflict."

Sue Groves, an active member of the Women's Support Group, worked as the bookkeeper for Cope Supermarket for eighteen years. Within days of the lockout, Sue found herself cashing the checks of the replacement workers who had taken her husband's job and forced to listen to their comments about the plant and the locked-out workers. Jeff Cope, the manager, was sympathetic, since both his father-in-law and his sister's father-in-law were locked out. Yet, according to Sue, his parents, who owned the store, "absolutely made it very plain that they were with the company," going so far as to cater events for companies such as Plasma Processing who were crossing the picket line to do business with RAC. Unable to bring herself to pay the bills for a company that had crossed the picket line, Groves quit her job. A deeply religious woman, Groves also lost her church, as she found herself unable to worship alongside managers and scabs.

The tensions reached down to the children as well, as hundreds of young replacement workers moved their families to town and enrolled their children in the schools. Although school administrators told teachers not to take sides in the lockout, some were married to RAC managers and salaried workers and made their company sympathies clear. Fights broke out in the schoolyard between children of Local 5668 members and children of replacement workers as they reflected the anger and bitterness they heard at home. During that first winter, after a meager Christmas and the long wait for unemployment checks, it was particularly difficult for the children of the locked-out workers to listen to the taunts: "I have new Nike shoes. You don't because your daddy isn't working and my daddy is. Your daddy's lazy. He doesn't want his job. My daddy works."

If all the scabs had been strangers, outsiders shipped in by RAC from far-away cities and towns to take their jobs, the situation might have been easier for the locked-out workers. The first busloads of scabs had been brought in from Cleveland, and others came from Florida, New York, Alabama, and Nevada. But not long after the lockout began, the company began to actively recruit replacement workers from the same communities where the union members lived, shopped, and worshiped. In many cases they were neighbors, in some cases they were relatives. Mostly they were young.

Trapped in low-paying service sector jobs, often struggling to support their families, they had been denied the American dream that their elders had found in the large aluminum, steel, and chemical plants that dotted the Ohio Valley. The jobs now available at Ravenswood were good jobs with good wages. They represented for many of these young men and women perhaps their first and only chance for economic security in an area where unemployment rates were twice the national average. Most knew that crossing picket lines was wrong. One couldn't grow up so close to the coal fields and not know that. But somehow it was not wrong enough to stop them from crossing the line at Ravenswood Aluminum. Terry Ashworth became a scab early on in the lockout even though his father was out on the picket line. He argued, "I lived around here all my life. It's just a decision I made to go in. You work for nothing all your life, you get tired of it. No insurance, no good checks, no overtime."

Few of the scabs had ever worked in a manufacturing facility before, much less an aluminum plant. But RAC was desperate to staff the plant, even if it meant replacing a very mature, stable, and skilled workforce with inexperienced, untrained, and in many cases unreliable employees. The first group they hired included a group of circus performers from Florida. At the December unemployment hearings RAC reported that the company had hired more than six thousand replacements in just the first six weeks of the lockout, yet was never able to keep more than a thousand in the plant at any one time. Dan Stidham describes the replacements: "They were all ruffians, really, and they created a lot of problems. . . . People was coming and going for whatever reasons. Some of them were getting fired. I guess they felt people just couldn't perform. It was a different kind of lifestyle. They weren't used to working in a plant. . . . They were just taking bodies, as long as they could get somebody in there to work."

Once those first terrible days locked inside the plant were over, the salaried and replacement workers had to commute back and forth to work each day past the bitter faces of their former friends and neighbors on the picket line. Working twelve-hour days behind boarded windows and barbed wire, those inside quickly developed a siege mentality. Their fears were fomented by repeated bulletins from RAC management warning them against union violence and cautioning them never to go anywhere alone nor to talk with anyone about what was happening inside. Management gave the salaried and replacement workers sweatshirts, jackets, hats, billfolds, and coffee mugs emblazoned with the RAC logo, and encouraged them to proudly display their pro-company status. Local 5668 members were clearly identified by their brightly colored union t-shirts or by the camouflage they wore. The two groups confronted each other daily on the sidewalks, streets, and playgrounds of Ravenswood, Ripley, and other small towns neighboring the plant.

Families were also torn by the lockout. Mike Bailes had to watch as his ex-wife crossed the line. "One [of my kids] was graduating and another one was a sophomore in high school. At that point in time, one of them was thinking of not going to college [because of the lockout]. But what's more difficult is the fact that I'm a union person and a negotiator, an official in the union. And her mother chose to cross the line, the very line that most stood for what I believe in."

Bailes's youngest daughter, who lived with her mother, came with him to the union hall and "she'd hear the word 'scab' and, of course, at that time she was only like four years old. And she . . . knew it was something bad . . . of course she didn't understand the whole significance of it."

Frank Ashworth, who had worked in the plant for more than thirty years and whose son Terry became a replacement worker, held the family together, but at great cost.

It hurt me that he went in, knowing we were out, but that's a decision that he had to make. You know, he's a man, not a puppet, and not a boy anymore, and he has to make his own decisions. It's something he knows he'll have to live with, as well as I do. And unions fought, not only in this plant, but in other plants all over the country to get what working people's got, so I don't know. I don't understand how people make a decision that the jobs are open, when they know that people are out on strike, trying to get a contract signed. But like I say, I don't make decisions for Terry, he makes his own. I don't live by his, and he don't live by mine.

For Local 5668 members, each of the more than seven hundred scabs hired that winter represented a deeply personal affront, a brutal violation of everything the union had worked for during the more than thirty years since the plant had opened. As the architect of the lockout, Boyle was the clearest enemy, but so were the increasing numbers of replacement workers who had taken their jobs, who were moving into their neighborhoods and conducting business in their towns.

As the aluminum kept rolling out of the plant, much of the locked-out workers' anger and frustration focused on tracking down, identifying, and confronting any and all replacement workers in their communities. They were joined in their efforts by the Women's Support Group and fellow Steelworkers from the towns up the river where Boyle and many of the scabs lived. For Gary Cochran, president of Steelworkers Local 5670 at Consolidated Aluminum, and his members, exposing and publicly chastising Boyle and his scabs became their mission.

I don't know if you can completely appreciate the war that went on up the river as a result of Ravenswood. You see, Boyle's supervisors and salaried people were sending their sons, daughters, relatives, and friends to Ravenswood to scab. Most of the communities around Ormet and Consolidated were small. There were a lot of Steelworkers and other union members living in these communities. Many times we knew the very first week that people began scabbing at Ravenswood, either from the union at Ravenswood or from people living in those communities. The exposure of the scabs began in December 1990. . . . At night leaflets were dropped on the streets, scabs' names were painted on rocks along highways, on barns, sheds, barn roofs, oil tanks, trailers, bridges, guard rails, and on the actual highway.

The graffiti were spray-painted at night, under cover of darkness, in large scrawling letters up and down the valley. Many targeted Emmett Boyle and RAC public relations representative Debbie Boger, particularly on the highways between Ravenswood and Boyle's home near Ormet. Boyle had been separated from his wife since 1987 and was in the middle of an ugly divorce. Everyone knew that Boyle had carried on a relationship with Boger for several years, starting when she was still married to one of the janitors at Consolidated Aluminum. The graffiti targeting Boyle and Boger, whom the workers had nicknamed "Rat-face" and "Sugar-booger," was extremely unsettling to Boyle. He would have his supervisors report to him each morning on any graffiti they sighted on the way to work and then send RAC security guards to paint them over, only to have the same message, or worse, sprout up somewhere else.

As the lockout progressed, more and more of the sign painting was done by members of the Women's Support Group. Tossing aside the soft pastels of their former lives as middle-class housewives, they dressed in black from head to toe, including ski masks, and slipped out at night to paint signs on rocks and barns. The work was scary and, if caught, they would face serious vandalism charges. But they knew that it was better that they face those charges than their husbands, who could be permanently fired for criminal activity.

The struggle was not just a war of painted words. With their futures at stake, some from both sides were bound to turn to more aggressive action. The union had worked hard from the beginning to avoid the violence that had escalated with the continued presence of the goon guards. They had purposely moved their picket stations away from the plant to avoid confrontation. They worked hard with their members to avoid the bloodshed that marred many strikes, such as the bitter clash between the UMWA and the A.T. Massey Coal Company in Elks Run, West Virginia, a few years earlier, where one scab was killed and many on both sides were injured, some seriously. Yet given the stakes, confrontations were impossible to avoid.

Within weeks of the lockout, the roads to and from the Ravenswood plant were strewn with jackrocks. Made from two long nails bent and welded together so that a point always faces upward, jackrocks have been used by striking Mine Workers to slow down traffic in and out of coal mines. As at Pittston Coal, the Steelworkers used jackrocks to slow down the trucks coming in and out of the plant and to sustain pressure on the salaried and replacement workers who were crossing the line each day. But in Ravenswood, jackrocks were used by both sides. While the union-made jackrocks were crudely fashioned from welded nails, replacement workers sometimes appeared to have more sophisticated jackrocks at their disposal, welded from two pieces of bent "rebar," with sharp chisel-like blades pounded at each end. Many from the union were certain that these jackrocks were being produced and distributed within the plant itself.

During the night, jackrocks were strewn on streets, shoulders, and driveways, taking a toll on the tires and windshields of locked-out workers and scabs alike. They were also thrown from passing cars, in some cases injuring motorists, pedestrians, and picketers.

In the first few months of the lockout, the local newspapers were filled with charges and countercharges of violence on the picket line and elsewhere. But most of the violence resulted in punctured tires and broken windshields. In late January, RAC charged the union in federal court with 162 instances of violence and vandalism between December 17 and January 23. Twenty-four of these were damage to vehicle bodies, 10 were damaged windshields, and 124 involved ruined tires.

Not all the incidents were minor. On December 23, a RAC guard lost his eye when a bullet shattered the windshield of his security van. No one was ever arrested for the incident. Early in the lockout, shots were fired from a car passing the union picket shack, narrowly missing members inside. Several arrests were made for vandalism and harassment on both sides, but few serious charges were filed.

One case that did go to court involved union picket captain James Picarella. Known for both his temper and his militancy, Picarella had had repeated run-ins with RAC management in the months before the lockout. In late February 1991 Picarella was pulled over by the county sheriff's department and later charged with carrying a concealed weapon and a loaded gun, as well as two coffee cans of jackrocks, two hundred ball bearings, a bag full of marbles, a wrist sling shot, a ski mask, and a pair of gloves. The charges against Picarella came shortly after he had spoken out publicly about preferential police treatment given to Ripley councilman and police commissioner Dave Brubaker, who worked as a RAC security guard, and Larry Glen, council recorder, who was working as a trucking contractor for RAC.

For the union, the violence was often framed in terms of that first march of the "goon guards" in full riot gear in late November 1990. In the weeks after the march, workers reported many cases of threats, harassment, and intimidation by the guards. On January 23, for example, a guard assaulted Vester Walker, who was on picket duty at the RAC north entrance. The replacement workers were also involved in this campaign, hurling jackrocks and intimidating workers. Marge Flanigan and several locked-out workers and family members were run off the highway by a man in a gray Pinto, wielding a gun. On February 9 Local 5668 member Buddy Haught stopped his car to check his tires, and a car stopped behind him. The driver pulled out a gun and threatened Haught.

Despite repeated company charges of union violence, the most serious incident during the entire lockout involved an attack on the home and family of a union member. In early January, Janice Crawford recalls,

I had carpal tunnel surgery done on January 7 . . . and I came home from the hospital about eight o'clock that night. And I was up and down all night long with my hand. My sister had two small children and her husband was out of town the night before and she came to my house to stay. She was afraid to stay alone. So I had my sister and my two nephews and my two daughters there. And [my daughter's boyfriend], he came to the door the next morning about seven-thirty in the morning and he said, "Geez, what happened to your car?" . . . And I looked out the door. The side of my car was all black and there was a rope coming over my car into my gas tank and they'd set it off, burned it and tried to bomb my car. That was the most terrifying thing I think I've ever had in my life. Because . . . my oldest daughter's bedroom was right beside the garage, the other daughter was above it.

The fire marshal explained to Crawford that her family was saved because her car's gas tank was designed to thwart such explosions. If it had been an older car, her family would have been blown to pieces. However, unlike incidents of purported union violence which were widely publicized and subject to intense police investigation, this attempted bombing received little attention in the media or from the local authorities. In fact, some in the press actually questioned whether it was strike-related, despite the fact that Crawford was identified as a union activist from her work at the Assistance Center and her car and home were plastered with union bumper stickers and signs. For Crawford and other Local 5668 members, this attack represented how far Emmett Boyle would go in destroying the their communities.

Where the official response to the attempted bombing of Janice Crawford's home was lukewarm, at best, an altogether different reaction occurred when, in the first week of March, two "fragmentation grenades," or pipe bombs, exploded near the homes of RAC employees. The first incident occurred early Sunday morning, March 3, 1991, when Thomas Redman, the father of a replacement worker, heard a loud blast in the night and woke to a large hole in the front yard of his home in Jackson County. The following day, Wood County resident Eddie Piggot, a former union member who had crossed the line and returned to work at RAC just two weeks after the lockout began, was talking with a neighbor in his front yard when objects thrown from a passing pickup truck exploded next to them, spewing an assortment of pennies, marbles, jackrocks, shotgun shells, and other fragments across the lawn and slightly injuring the visiting neighbor.

Telling the media, "As soon as you make a bomb, it's federal," the local sheriff immediately brought in the FBI and the Bureau of Alcohol, Tobacco, and Firearms of the Treasury Department to investigate. Within the month two young locked-out workers, Robert Buck II and Gerald Church Jr., were indicted in federal court for manufacture and possession of unregistered firearms.

Ed Piggot, the target of the second pipe bomb, had been Gerald Church's best man and close friend before the lockout. Ever since Piggot crossed the line, Church had regularly enlisted Buck and others to drive by the Piggot house and throw jackrocks and paint bombs. Within weeks of the indictment, Church had turned state's evidence, agreeing to testify against Buck in return for a lesser charge. Buck, twenty-six years old, with no criminal record, suddenly faced up to twenty-five years in prison and a $750,000 fine.

Given the polarization of the community and the long tradition of strike violence in the West Virginia coal fields, the Ravenswood lockout actually stands out for its lack of strike violence. This was no accident. The local leadership and the international union were fully aware that Pittston Coal had effectively used the courts to wage a war of injunctions, judgments, and fines against the Mine Workers union. These judgments not only severely restricted picket line activity but also assessed penalties in the millions of dollars through which Pittston intended to bankrupt the UMWA.

From the beginning, the local leaders publicly stressed the importance of avoiding violence. After Judge Fox issued the initial injunction in early

December restricting the number of pickets, both Stidham and Chapman repeatedly cautioned members in meetings and newsletters to avoid doing anything to invite further injunctions:

> The union has stressed, from the outset of our lockout by Ravenswood Aluminum Corporation, that there was to be no picket line misconduct. We have a legal right to peacefully picket and inform the public of the labor dispute in which we're involved. However, acts of violence or misconduct of any kind are not legal and therefore cannot be tolerated by the union.
>
> As you may know there have been some incidents where automobiles were damaged, tires flattened, and persons assaulted. The union has no knowledge of who may be responsible for such acts. However, we do know that IT MUST STOP! Such acts will not help in our efforts to negotiate a fair contract, and may, in fact, do us harm by having our pickets further limited by either the courts or the National Labor Relations Board.

The local kept busy staying on top of the lockout. The union hall was staffed twenty-four hours a day, seven days a week. All phone calls, visitors, and information were recorded in log books kept on Dan Stidham's desk. A radio room was set up in the rear of the hall, to monitor all communications inside the plant; these were recorded in another set of log books. The local also stationed members at a "river shack" across the Ohio River from the plant, as well as staffing the picket lines along Route 2.

The injunction limited the union to six pickets at each of the picket stations and, with the continued presence of the goon guards and an increasing number of scabs, a feeling of impotence was growing. Given the stakes, the situation was ripe for violence to erupt. But the local took an entirely different direction with an action organized by the Women's Support Group.

"On January 28 we wanted to do something to let the company know that we were not going to stand by and watch it happen," reports Marge Flanigan. Shortly after 5:00 A.M., the women assembled and spent the entire day driving slowly in front of the plant, completely snarling traffic in and out. They repeated it the next day. "And then on the second or third day we got 265 tickets," Women's Support Group member Linda McCoy reports. "The third day the same amount of tickets, and then the fourth day we'd just take a ticket at random. We told them we already had a ticket. We didn't care if they ticketed us to death, we were not stopping."

USWA District Director Jim Bowen told the women, "Don't pay them.
And you go over to that damn magistrate's and you tell them you want
a jury hearing, every damn one of you. Pretty soon the state police will
say, 'Jesus Christ, we'll be over there all day long.'"

The actions came to be known as "drive-bys." The police responded
by pulling out individual cars and putting them in a "penalty box," as if
they were playing a hockey game. Jim Bowen says, "They finally decided
we weren't going to pay [the tickets] and they put the penalty box in. I
thought it was hilarious. 'You're penalized for one lap!'" This only made
the drive-bys more fun. Some of the men were jealous because injunctions prohibited them from participating in the drive-bys, so some
women put wigs on their husbands and took them along for the ride.

According to the salaried employees, the drive-bys had a tremendous
impact inside the plant. "The idea of having to confront the wives of the
locked-out workers and the traffic was daunting. Management responded, 'Oh, my god, they're back, they're back,' and everybody
wanted to go home," reported a worker who was inside at the time.

It was not only the drive-bys that rattled management. "Charlie Mc-
Dowell's vehicle tore them up. If they would see his vehicle coming
down Route 2, it was alerted all over the plant. . . . Oh, my God, you'd
think we were going to be raided . . . they'd come down the hall, they're
yelling, 'Here comes Charlie, here comes Charlie.'"

Throughout February and early March the drive-bys continued and
were an important focus of the local's activity. On March 23, Judge
Haden issued a restraining order prohibiting any further drive-bys. The
injunction might have come just in time, before the drive-bys escalated
into something beyond the local's control. As Stidham describes it, by
the end of March the drive-bys had gained a momentum of their own.

> The drive-bys were getting bigger. . . . People were looking forward to it
> more and more every day. We were getting outsiders coming in and par-
> ticipating. Things were getting out of control. I know that there were fed-
> eral marshals in those drive-throughs, mixing in with our people and the
> scabs, and you didn't know who was who. And there were some people
> that were just out there for whatever, they were having a good time. It was
> something to do. It took their minds off their real problems. But it could
> have created some real problems . . . You know, someone from outside
> coming and taking over, guns drawn, maybe shots fired, some beatings,
> people dragged out of their cars, busted windshields. All that kind of
> stuff, and it would just get worse and worse. Judge Haden finally put an
> injunction against us on that, and it probably was for the best.

The drive-bys were important to the union. They solidified the Women's Support Group and silenced any doubters about the women's place in the struggle. And in that first gray winter, the drive-bys offered the locked-out workers a way to laugh in the face of management's arrogance. But the drive-bys, like the jackrocks and graffiti, had great limitations as strategy. Beyond letting off steam, they did nothing to leverage RAC back to the bargaining table or bring the workers back into the plant.

8

A SECOND CHANCE

Throughout the winter the union bargaining team continued to meet with management and the mediator in court-ordered bargaining. Although the union had revised and dropped many proposals in an attempt to reach a settlement, the company refused to budge. Instead, RAC management dug in their heels by hiring more scabs and filing more legal actions against the union.

On January 9, 1991, RAC manager Don Worlledge and company spokesperson Debbie Boger held a press conference inside the boarded-up administration building. They announced a new RAC Scholarship Fund and reported that through the "outstanding performance" of the replacement employees the company had broken a number of pre-lockout production records in both output per man-hour and output per machine. According to Worlledge, "I don't know whether it's higher motivation or what, but they [the replacement workers] are outperforming our previous workforce. They are not fettered by work rules. There is none of this 'it's not my job' talk. It is obvious we can do what we have been doing with a lot less people."

Worlledge and Boger stated unequivocally that all 725 replacement workers were now considered permanent employees and that the company would not negotiate with the union over their status. As far as RAC was concerned, even if a settlement was reached, at least 700 union workers would be left out in the cold. In Boger's words, "It is our intention that the permanent workers will stay. Yes, a two-month worker could edge out one of the union workers." She added that the company was committed to hiring even more replacement workers, which would further reduce the number of jobs available to union members if they were ever to return.

Stidham's reply was swift and to the point: "If we don't all go back, none of us go back. None of us will ever work with a scab and we don't intend for any scab to take one of our jobs. I don't think we could ever reach an agreement where the scabs stay in. We have put our blood, sweat, and tears into that plant and some of us have given our lives. We helped build the plant up and have worked there for thirty-plus years to make the contract what it was. We don't intend to let someone come in off the street and reap those benefits."

The nadir came just a few days later, on January 17, when the bargaining team was in Pittsburgh for a negotiation session called by federal mediator Carmen Newell. Because the last union proposal had included significant economic concessions, the union leaders came expecting a counteroffer from RAC. But, like so many bargaining sessions before, this one lasted only ten minutes, just long enough for Schick to tell the union that the company no longer had any offers on the table. "Previous offers included deadlines, which have long since expired."

Making matters worse, after the session was over, Joe Chapman received word from the NLRB that the union's unfair labor practice charge, filed right after the lockout, would be dismissed. The union had been depending on the charge to establish legally either that RAC's actions constituted an illegal lockout or that the union was engaged in an unfair labor practices strike. By law, such a ruling would define replacements as temporary and could result in the locked-out workers' eventually getting their jobs back, with full back pay. If the charge was dismissed, the lockout would be labeled an economic strike and RAC would be free to make the scabs permanent.

With a very long face, Joe Chapman wandered into USWA staff attorney Paul Whitehead's office at the international's legal department that January morning. After talking over the situation for a few minutes, Whitehead decided to call the NLRB.

I reached the guy and he said, "Yeah, what Joe told you is true." And I said, "Gee, that's disappointing. . . . Is there any chance of our being able to submit some additional information?" And he said, "I don't know. . . . I think the letter's out." And I said, "What does 'out' mean?" And he said, "Well, it could be in the mail room in the out box." And I said, "Could you go look?" And he . . . put the phone down, and took a long time, and said, "It's in the regional director's out box, and it'll be mailed at mail time." This was about one o'clock in the afternoon. . . . And I said, "Okay, hold that, please. Let me talk to you in a little while."

Whitehead then went into the library and asked Rich Brean, the Steelworkers' resident expert on unfair labor practices, if he could work on this case.

[Rich] said, "Yeah, I'll throw myself into it." And, knowing that, I went back to talk to the Board agent and he said, "Okay, if you want to submit some additional affidavits, I'll go to the mail room and pull it out of the out box." And he did that. The company, of course, had received a phone call [that the case was going to be dismissed]. They were probably delighted. They never did receive that letter giving the official word.

The union's original charge, filed by District 23 counsel Stan Hostler, had been a standard one-line "bad faith bargaining" charge, alleging that RAC had bargained in bad faith by unilaterally implementing changes in wages, hours, and working conditions after falsely declaring that impasse had been reached. In stopping the letter from going out, Brean and Whitehead had pulled the union's entire unfair labor practice case out of the fire, with minutes to spare. The union could have appealed a negative decision, but it would have been a long uphill battle while RAC's victory devastated the locked-out workers.

Even though the NLRB had not ruled in the company's favor, RAC still claimed victory. Earl Schick told reporters that the only reason the union withdrew its case was that the NLRB had informed both sides that the charges would be dismissed. "This basically clears the air of any union allegations claiming that RAC has bargained improperly," Schick said. "It also effectively affirms the status of RAC's present hourly force. . . . Even if the union files additional charges based on subsequent incidents, the NLRB's decision on this impasse issue will not change. We have maintained from the beginning that we acted lawfully in this matter. This is a big step in returning to normal operations." Schick's claims were reinforced when the acting regional NLRB director, James Ferree, reported to the press that the NLRB planned to dismiss the union's charge because "we don't think we could convince an administrative law judge that there was a violation here."

But the union strategists had not given up on the legal challenge. They had bought themselves a second chance. Chapman told the press that the union had decided to withdraw the charges in the aftermath of the January 16, 1991, bargaining session, in order to consolidate the earlier charges with more recent unfair labor practice violations and put before the NLRB "the whole sorry record" of RAC's conduct.

Meanwhile, Rich Brean threw himself into the case, as he had promised. His first step was to go to West Virginia to interview Joe Chapman and put together a new unfair labor practice charge. For two days he sat with Chapman and took copious notes on long legal pads, asking Chapman to just "tell him everything." Initially, Chapman was not impressed. From behind his thick glasses, Brean came across to Joe more like a rumpled, absent-minded professor than like a hot-shot labor attorney. After a few hours of questioning, Joe stepped out and told his secretary, "Jesus, you should see that lulu they've sent us here this time. He ain't doing nothing. Hell, I might as well wrote this up and sent it in to him."

The next day, after Brean had returned to Pittsburgh, he faxed Chapman an affidavit to sign. "And that was like forty-nine pages or whatever he faxed down here and I couldn't believe, I had to make very few changes. I mean he had sat there just with that little old scribbling every so often, and he wrote it like a storybook. . . . I thought he had really done an excellent job."

With Richard Brean in his corner, Joe Chapman, for the first time since the lockout began, did not feel quite so alone. Brean had graduated from Harvard Law School, class of 1978, where he had been a classmate of Paul Whitehead, now also on the Steelworkers legal staff. His mother had come from a working-class Jewish family with a strong union tradition. Yet, unlike Whitehead, who had majored in labor studies at the University of Wisconsin and always knew he wanted to work for the labor movement, Brean took an internship with a management law firm. He quickly realized that he loved labor law but did not like working for management. "I asked for a suggestion from Leonard Sheinholds, who's chairman of their labor group, about where to work on the union side in Pittsburgh. He said basically there was only the Steelworkers legal department, and he called up and made the first contact for me, Carl Frankel, and said, 'There is a misguided young man here.' And I walked down that day, that's how I got into the union."

Sixteen years later Brean had become the USWA expert on unfair labor practices, strikes, and lockouts. As he sat there in Charleston listening to Chapman tell his gloomy story, he was barely able to conceal his excitement. He knew he had a winning case. He just couldn't believe that the case hadn't been put together before, nor how close the local had come to losing everything.

The original union unfair labor practice charge had focused exclusively on impasse. An expert on lockouts, Brean knew that whether the parties were at impasse was much less important than whether RAC had bargained in good faith. Since 1965, based on an NLRB case called *American Shipbuilding*, employers have been able to legally lock out their employees and hire temporary replacements. They don't have to be at impasse, and they don't have to prove that the union is about to go on strike or sabotage equipment. Basically, a lockout is the employer's version of a strike: their way of putting pressure on the union to agree to the company's demands. Brean explains, "[Lockouts] are perfectly lawful [when] the employer has bargained in scrupulous good faith before they lock out. . . . And that's why [a lockout is] actually more dangerous for an employer than a strike [is] for a union. If you're a good union lawyer, many, many, many lockouts are tainted because there isn't good faith bargaining."

As he listened to Chapman, Brean realized that the Ravenswood lockout was clearly tainted (made unlawful) by bad faith bargaining, both before the lockout occurred and in the months that followed. It was further tainted by management's unilateral implementation of the final offer one month into the lockout, without reaching impasse. Most important, as long as the tainted lockout continued, the company was accruing back pay liability, because once the lockout became unlawful the employer lost its legal right to hire replacements. For Brean, this meant that the Steelworkers not only had a strong case, but, if they won, they could win big.

With this in mind, the new charge that Brean put together contained three major elements. First, RAC had engaged in bad faith bargaining by refusing to discuss pot room safety issues, refusing to respond to information requests from the union, prematurely and illegally declaring impasse, and implementing its final offer before impasse was reached. Second, this bad faith bargaining tainted the lockout, making it unlawful. Third, the employer committed yet another unfair labor practice when it permanently replaced the bargaining unit employees who had been unlawfully locked out.

On January 28, seventeen days after the original charge was withdrawn, Rich Brean filed the new charges at the Cincinnati regional office of the NLRB. The Cincinnati region did not have a pro-union reputation. In the earlier charge, the union had had to deal with a Board agent "whose main thing was that he wanted to get back to Cincinnati before it got dark." Under NLRB procedures, if the Board finds that the charging party's case has merit, the Board agent then acts as an advocate for that side. The agent presses for a settlement and, if necessary, takes the offending party to trial before an administrative law judge. Thus the choice of an agent was critical for the union. This time, to Brean's delight, the union was assigned an agent named Carol Shore.

Shore was considered the best agent in the region, "really smart and hard working." Brean's task was to convince her that RAC had locked out its workers and that the union's case was worth fighting for. First, he called her on the phone and spoke with her for an hour and a half. By the end of the conversation, Shore was on board but her supervisor was still skeptical. So Brean went to Cincinnati to present his case in person.

I went down and sort of convinced them in an oral presentation that the charge had merit . . . because we had this argument that the lockout never ended. See, the safe way to do this, and the good fallback argument, is that there was an unlawful lockout for one month and that was converted into an unfair labor practice strike. And, at the very worst, at least our people can't be replaced. And that was the safe way to fight the thing. [But] I really thought we needed more. That was conventional warfare. I thought we needed nuclear weapons, because these people [RAC] were animals. And I really thought that the only way we'd get Boyle's attention and the company's attention was to try and create this mega-liability. And that's when we cooked up the idea of the lockout never ending.

Brean felt fairly certain that he had won over both Shore and her supervisor to his side. Several months would pass before the union would hear back from the Board on whether they were going to issue a complaint but, at least for now, the NLRB case had been put back on track.

9

JUMP-START

The close call with the NLRB case reinforced Joe Chapman's desire to get the international union more involved in the campaign. Although the local union members were holding strong, the company had brought in close to a thousand scabs, and one way or another they were producing aluminum. Charlie McDowell's prediction that scabs could not possibly turn out a quality product was correct, as shown by the scrap carried out of the plant and the large number of orders returned by dissatisfied customers. Yet RAC and Boyle seemed determined to get production right and prove the union and McDowell wrong.

Several weeks after the lockout began, Chapman had approached USWA president Lynn Williams and vice-president George Becker about becoming directly involved. But at that point Williams and Becker did not share Chapman's sense of urgency. As the principal officers of an international union with 500,000 members, they could not get involved in every local labor dispute, particularly if it appeared the dispute could be resolved quickly through normal bargaining channels.

By the beginning of the year the pressure was starting to build for Chapman, who had day-to-day responsibility for the union, its members, and their families. For Joe, doing everything he could to win this lockout had become a deeply personal struggle. At times he felt as if he was carrying seventeen hundred people on his shoulders: "I built a new home on what was bought by my great-grandfather in 1838. I don't intend to go anywhere else to live. So I know a lot of those folks that work there. . . . You know, in West Virginia, you're a neighbor if you're within ten miles. . . . I just couldn't visualize going the rest of my life living around a community that I lost a big local at. And I was going to make sure . . . that my conscience was clear that I'd done everything I could."

One of Chapman's main contacts at the USWA international offices in Pittsburgh was communications director Gary Hubbard, who had helped the local deal with the press since the beginning of the lockout. Joe persuaded Hubbard to speak to Williams about becoming more involved. He felt that, if help did not arrive soon, the plant and seventeen hundred union members would be lost forever. With the lockout in its third month, Williams was more receptive; clearly normal channels were not working. Fearing that a loss at Ravenswood could have a devastating effect on workers throughout the aluminum industry, Williams asked Becker to become directly involved in the lockout.

In the USWA structure, Becker was responsible for bargaining the large master aluminum contracts. But more important, he had worked closely and successfully with Emmett Boyle on two other occasions. The first contact had come out of nowhere. "I got a call from [District 36 director] Thurman Phillips. 'George,' he says, 'I been dealing with a guy by the name of Emmett Boyle who is trying to open up the Burnside plant [in Louisiana]. The son of a bitch may call you but I want you to know that we've done everything we can do, we just can't deal with him.' And I said, 'Well, what makes you think he'll call me?'" Phillips explained that Boyle wanted to talk to somebody in the international who knew something about aluminum, "that could put together a deal." Becker continues,

> Subsequently I got a call from Emmett Boyle. . . . He told me about having such a wonderful contract with Ormet, how he's dealt with the Steelworkers for years, he respects the Steelworkers. And he has a chance to open up this plant, put a lot of people to work, and he also has a lot of caustic soda. And he has this caustic soda in barges, that's all along the coast and other places. I said, 'Where did you get all of this?' He said, 'I just got it.' And I mean it just started sounding a little bit strange. And it takes a lot of caustic soda in order to be able, through an electrolytic action, to process the bauxite into alumina and there's a worldwide shortage of this caustic soda. But he's supposed to be operating on a shoestring and yet he's got all these barges of this stuff sitting out there. It didn't sound right.

Becker met with Boyle. "We eventually wound up with a good, decent contract, a Steelworker contract," Becker recalls. A bit later Bowen asked Becker for help in some negotiating difficulties at Boyle's Ormet plant. "I did so, apparently with Boyle's appreciation. I was a hero of sorts in Boyle's eyes at that time."

So Lynn Williams had every reason to believe that Becker might be able to remedy the lockout quickly. "Two times up at bat and two times a home run, okay," recalls Becker. "So that's why Lynn thought, 'Hey, if Emmett Boyle's the bastard in this thing, if he's the fly in the ointment, then at least George can talk to him, maybe even deal with the guy.'"

Becker began by investigating the lockout in detail. It did not look good. He then traveled to West Virginia to assess the situation in person. What he found was sobering:

> I found that our people were sitting on a picket line with no effective means of putting the company under pressure whatsoever. Our members couldn't even picket in front of the plant. They were under an injunction. They could stand at the side of the road a mile from the gate and that was it. And I found at the time we had about eleven hundred scabs in the plant already. No way to slow them down. No way to stop them. No way to put pressure on them. I found that we had a lot of people fired [permanently replaced], and people were generally afraid to do anything and were not doing anything. Simply said, nothing was being done that would be effective in pressuring the company to return to the bargaining table.

Upon returning to Pittsburgh, Becker decided not to call Boyle just yet. "I don't want to talk to a person that's in the key position of settling with me without having some strength on my side. We had no strength, zero, when I looked at it. We didn't even have enough strength to do other than just beg, and we don't do that."

But Becker was not about to give up. It was not in his nature. He had never forgotten where he came from.

> I've been a steelworker since I was fifteen years old, in Granite City, Illinois, Granite City Steel. I went to work on summer vacations. This was during the latter few years of World War Two. After the second year I stayed, then went into the military service right at the end of World War Two. Came back and went back into Granite City Steel. Worked in other steel mills and foundries in that area of the country and then went back in the service during the Korean War. When I came back I went into the same job at the mill that I had when I was fifteen years old.
>
> Well, if you worked in the mill where I worked and the age I went in, the fact that you were physically there made you pro-union and a strong advocate of worker representation of some kind. Hard, physical labor, dirty, dangerous, this mass of humanity and a lot of injuries, zero dignity, not any consideration by the company at all. It was a rough existence. It was easy to become active in the union and it was only to the degree of militancy you had and the leadership that you were able to cultivate along the way that determined whether or not you were going to go anywhere with it.

Becker held on to his working class roots and was most at home with rank-and-file workers. A no-nonsense family man, he was not one of those union officials who reveled in lunches at swanky restaurants or fancy cocktail parties. Trim and handsome, Becker had the chiseled features of a much younger man. He had stayed in excellent shape since his days in the Marines and always seemed ready to fight.

Although Becker's assessment was that the situation at Ravenswood looked grim, he saw some positive signs. Most significant, despite the scabs, the local was still unified. Less than one percent of the workers had crossed the picket line. Membership solidarity was essential, though not sufficient, for a union victory. While Becker may have felt earlier that a call to Boyle could have reopened bargaining, he could now see clearly that much more than a phone call would be necessary. According to Becker, "Like in most things, there's a lot that can be done if we have the determination and the resources to be able to do it, and capable, dedicated people."

Within a month Becker brought together a number of key union staffers—Bowen, Chapman, Hubbard, and Bernie Hostein, director of organizing—plus Dan Stidham. At that first meeting, it was clear that this was not going to be an easy fight. Everyone in the room knew—Emmett Boyle was out to bust the union. To win would take much more than picketing a mile from the gate and filing charges with the labor board. So Becker assembled a strategy team of the best and brightest inside and outside the Steelworkers' organization to put together a campaign against RAC.

Becker was well aware that, since the 1970s, when American employers took off the gloves in their fight to break the power of organized labor, unions had been exploring alternative channels to bring pressure on employers, beyond the limited protections provided under labor law. To employers set on staying nonunion or breaking existing unions, unfair labor practice charges and bargaining orders amounted to little more than annoying pieces of paper. The strike in particular had become a much less viable weapon, as employer after employer forced unions to strike, hired permanent replacements, and a year later held decertification elections to rid themselves of the union.

One of the early highly visible efforts to move beyond traditional union strategies was the struggle by the Textile Workers Union in the late 1970s to organize J.P. Stevens. Notoriously anti-labor from its base in the deep South, Stevens had proven impervious to traditional organizing techniques for more than two decades. Under the direction of staffer Ray Rogers, the union turned its attention away from the southern communities themselves to the J.P. Stevens board of directors. The goal of this "corporate campaign," as Rogers dubbed his strategy, was to pressure board members not just in their roles as Stevens directors but to extend the campaign to the roles they played in other organizations.

For example, Rogers organized a highly successful campaign with women who used Avon cosmetics and jewelry. He connected Avon chairman David W. Mitchell, who also happened to be a Stevens board member, with J. P. Stevens' aggressive opposition to its women employees' right to organize. Ultimately, according to *Newsweek,* "Avon Products chairman David W. Mitchell resigned from the Stevens board after receiving a rush of mail and telephone calls questioning his affiliation with a company that has long been organized labor's number one pariah." Mitchell, along with Stevens chairman James Finley, also resigned from the Manufacturers Hanover Trust Company board after a number of unions threatened to remove their pension funds. As attacks on directors at New York Life Insurance Company and other financial institutions connected with Stevens continued, the company finally relented and signed a union contract in 1980.

In the aftermath of the Stevens victory many union activists were enthusiastic about using corporate campaigns. Longtime *New York Times* labor reporter Abe Raskin wrote, "Under the leadership of an activist whose name few businessmen have heard of, organized labor is bringing some of the country's biggest corporations to their knees by substituting a money whip for its blunted traditional weapons of strikes and mass picketing."

Yet the victory at Stevens did not quickly translate into other victories for the labor movement. The Steelworkers began a fledgling corporate campaign during the Phelps-Dodge strike, and the Machinists started one during their bitter strike at Browne and Sharp in Rhode Island. Both failed. These corporate campaigns were not a good test of the tactic, as they were last-minute add-ons to strikes well under way. Still, it appeared that labor's new weapon might be less effective than many had hoped.

In 1984 Ray Rogers, now a union consultant, became involved with the struggle of fifteen hundred workers in the small town of Austin, Minnesota, on strike against the Hormel meatpacking company. Through a corporate campaign, Rogers gained national media attention for the strike of Local P-9 of the United Food and Commercial Workers (UFCW). But in a struggle that polarized the American labor movement, the international withdrew its support of the local union. Although the local and Rogers continued the campaign for many long months, they were unable to bring Hormel back to the table. The international union ultimately put the local union in trusteeship.

While the loss at Hormel did not prove or disprove the efficacy of corporate campaigns, it seemed that Rogers's quick fix for the labor movement was not sure-fire. A lost strike by the Paperworkers union against

International Paper Co. in 1988, which also included a corporate campaign led by Rogers, raised further questions.

However, corporate campaigns targeting boards of directors were not the only creative tactics tested by the labor movement during the 1980s. In 1982 in New Bedford, Massachusetts, Local 277 of the United Electrical Workers (UE) went on strike against corporate giant Gulf and Western, which had bought the locally based Morse Cutting Tool company. Galvanizing the entire community around Gulf and Western's disinvestment policy, the union won a thirteen-week strike.

In 1981 in St. Louis, United Auto Workers (UAW) staffer Jerry Tucker helped orchestrate an "in-plant strategy" designed to pressure Moog Automotive, Inc., back to the bargaining table while the workers stayed at their jobs, working without a contract. At contract expiration, Moog had presented the union with a final offer that included a wage cut of more than $3 an hour. Instead of going on strike and permitting Moog to hire permanent replacements from the legions of the unemployed, UAW members stayed inside: working to rule, refusing overtime, filing mass grievances, and holding regular noontime and break-time rallies. Six months later, a united membership emerged victorious with a 36 percent increase in wages and benefits.

Similar inside campaigns were used by the Steelworkers at Cascade Rolling Mills in 1985, the UAW at a Schwitzer cooling fan plant in 1983, and by the Service Employees International Union (SEIU) and Local 1199 at major hospitals in Boston and Philadelphia. While corporate campaigns, as conceived by Rogers, targeted decision makers and their allies at the top levels of the corporate hierarchy, the in-plant strategy used direct action by workers on the shop floor and in the broader community. This strategy also protected workers from the discharges and decertifications that resulted from more traditional strikes.

Labor's experience in the mid-1980s showed that no individual technique, not corporate campaigns nor inside tactics, was the panacea for an embattled union movement. Unions found that even with full community support and a united workforce, many employers were impervious to community-based campaigns. Too often in the global economy, decisions affecting workers' lives were being made by financiers and dealmakers thousands of miles away. Thus a consensus began to develop by the late 1980s that none of these strategies alone was sufficient to win. Clearly, successful campaigns, such as the UMWA campaign at Pittston Coal, the Communications Workers' and International Brotherhood of Electrical Workers' campaign at NYNEX, and 1199's campaign with the New York League of Voluntary Hospitals, depended on running a multi-faceted campaign that put escalating pressure on the employer in the workplace, in the community, in the corporate boardroom, and around the world. Hence the term "corporate campaign" was replaced by "coordinated campaign" or "strategic campaign."

Hoping to build on the wealth of expertise gained in strategic contract campaigns at Pittston, NYNEX, the *New York Daily News,* and Eastern Airlines, Becker decided to go outside the Steelworkers for research assistance in putting together the full-scale coordinated campaign he had in mind. He originally planned to work with the Food and Allied Service Trades (FAST) Department of the AFL-CIO, which had extensive experience in corporate research and in developing campaigns. But the Steelworkers were not members of FAST. While Becker was waiting to hear from Lynn Williams about joining FAST, he decided to contact Joe Uehlein, director of special projects for the Industrial Union Department (IUD) of the AFL-CIO.

The USWA was a key union in the IUD, and Uehlein had come out of the Steelworkers and was the son of a longtime Steelworkers staff member from Pennsylvania. Uehlein and the IUD had a great deal of experience assisting unions with coordinated campaigns, having worked on the Eastern, Carrier, and *Daily News* strikes. The IUD had also been involved with the Steelworkers in the fledgling campaign at Phelps-Dodge. After talking with Uehlein, Becker decided to work with the IUD rather than wait to join FAST.

Once on board, Uehlein put together an investigative team including Richard Yeselson, a corporate researcher who had worked at FAST and most recently with SEIU's Justice for Janitors campaign, and Jim Hougan, an investigative journalist and private investigator who worked for law firms and unions, including the IUD.

The team first reviewed the materials that had been gathered in the local union hall. In December, RAC's Strategic Plan for 1990–1991 had been leaked to the local from an unknown source within the plant. The document, prepared by the prestigious accountants and business advisors Price Waterhouse, was a windfall for the union. While union leaders knew that Bradley, Boyle, and Strothotte were involved with RAC, their knowledge about RAC's ownership ended there. The union learned from the Price Waterhouse report that 20 percent of RAC was owned by Charles Bradley, whose Stamford, Connecticut-based firm, Stanwich Partners, had bought the plant from Maxxam/Kaiser. Emmett Boyle, who less than a decade before had been a lowly reduction plant manager, now owned 27 percent of the company.

The key player, however, was Willy Strothotte. Through a company called Rinoman Investments, Strothotte owned 48 percent of RAC. Through another company named Ridgeway, Strothotte also owned all of RAC's preferred stock and held $54.75 million of the company's debt. Steelworkers at the local and the international had never heard of either of these two companies, but the Price Waterhouse document described them as affiliates of the Clarendon corporation.

Clarendon, like Stanwich Partners headquartered in Stamford, was a major international player in the aluminum industry. District director Bowen knew that Clarendon had been involved as a supplier and as a customer to Ravenswood and Ormet, but did not realize how deep the connection ran. In addition to being a major stockholder and "banker" in terms of holding the debt on RAC, Clarendon was slated to take over providing alumina to RAC when RAC's contract with Kaiser expired. Indicating an even more nefarious relationship, the Price Waterhouse report noted that, once potline number one was restarted, all metal would be sold directly to Clarendon. Clarendon, through a paper transfer, would then sell the metal back to RAC for use in the Ravenswood plant, at a profit to Clarendon.

This incestuous intertwining of RAC and Clarendon became public during those unproductive bargaining sessions at the end of January. RAC had long insisted on ending the workers' "metal price bonus," based on the market price of aluminum, and in its place was suggesting profit-sharing. After learning about the "hidden profit" being paid to Clarendon, Chapman was furious about the proposal. While he said the union would be happy to be involved in real profit-sharing, he told the press they were not interested in a "pig-in-a-poke" plan.

RAC reluctantly acknowledged the agreement with Clarendon, but denied that Clarendon was involved in any way in the ownership of the company. The company defended its action. "Mr. Chapman implies that he has found some type of unethical business practice in the document he has quoted. That is not true. In fact, under the original sale and purchase agreement there was an understanding that some of the metal produced at Ravenswood would be sold to Clarendon at Ravenswood's production costs plus a fixed guaranteed profit."

Management was furious that information gleaned from the Price Waterhouse document had been leaked to the press. They issued a stinging statement about the union's use of the report. "We have been aware that the union has in their possession a copy of a confidential document which, to our knowledge, could only have been obtained from someone who stole the document from confidential files. Stolen documents are rarely fully understood and the union simply does not understand this document or the business practice it concerns. The company has been investigating its legal options regarding the union's unauthorized possession and public use of this information."

But more important than the agreement to sell back RAC metal at a profit, the Clarendon connection confirmed that international financier Marc Rich was involved. Rich had originally been in on the bidding with Charles Hurwitz to buy the Ravenswood plant and, although he had dropped out at the last minute, he was rumored to have maintained some connection to the company. Everyone in the industry understood that Marc Rich controlled Clarendon.

10

PENETRATING THE VEIL

Marc Rich was the son of a low-level trader who fled from Frankfurt, Germany, to Antwerp, Belgium, in the late 1930s, after the rise of anti-Semitism. Born in Antwerp, Rich was raised in Philadelphia, Kansas City, and later New York, as his father's business grew and prospered. He did well in school, with the exception of math, but remained distant from his fellow students. In many ways he remained more a product of his European roots than of his American environment.

Rich enrolled in New York University in 1952 but, bored in college, dropped out two years later to work for Ludwig Jesselson at Philipp Brothers, one of the largest commodity trading companies in the world. At Philipp Brothers, Rich remained aloof. One of his colleagues remembers him from the mid-1950s as the "tall man with the soft voice and the strained smile who always wore Saks Fifth Avenue suits." Another colleague more pointedly recalls, "Marc always felt he was brighter than us, that his shit don't smell. And he never talked about anything else except business."

Yet his lack of camaraderie with his peers did not stand in the way of Rich's progress at Philipp Brothers. One of the senior officers in the firm remembers, "He had the best memory of anyone at the company, and the nature of a true gambler. He never stumbled into anything in his life. Every risk he took was calculated." As Jesselson's favorite young trader, Rich was sent to Cuba and Bolivia, and in 1967 he became manager of the Philipp Brothers office in Madrid.

In the early 1970s Rich became interested in trading in oil, which had long been off limits in the Philipp house. He developed important inside connections with the Iranian National Oil Company and began shipping crude oil to Spain. He linked up with Pincus (Pinky) Green, another Philipp trader, who had tremendous experience in the shipping business. According to Craig Copetas, author of *Metal Men: Marc Rich and the 10-Billion-Dollar Scam,* "Using Philipp Brothers' bank lines, Rich and Green embarked on oil deals without informing New York of their details. There were no hard and fast rules in trading oil, so they made them up as they went along."

While Rich delivered on his oil deals as he had done in metals and other commodities, the board of directors of Philipp Brothers grew concerned, given the volatility of oil and the power of OPEC. In their pursuit of oil, Rich and Green made a deal to purchase a Greek oil tanker, but the deal was nixed as soon as Jesselson got word of it. Amid these tensions about their oil trading, Rich and Green each demanded $400,000 bonuses from Jesselson. When Jesselson refused, Rich turned on his old mentor and left with Green and six of Philipp Brothers' top traders.

The new firm, Marc Rich & Co., A.G., was financed with a $2 million loan arranged by Rich's father and $1 million cash supplied by Jacques Hacheul, who followed Rich from Philipp, in cooperation with Iranian senator Ali Rezai. The company was headquartered in Zug, Switzerland, taking advantage of local tax laws that made Zug highly favorable for business. Starting in the 1960s and 1970s corporations began using a loophole in the 1946 Zug tax code that allowed them to pay a third less tax than in the rest of Switzerland. Corporations flocked to Zug, which boasts one corporation for every two inhabitants. Marc Rich, International, the U.S. affiliate, had offices in New York.

Quickly, Rich's new firm became a major player in the world oil market. One trader noted, "Rich had become so big in oil that he seemed to appear like a Saudi sheik wherever there was an oil deal to be made, often to the embarrassment of the American oil companies." By 1976 Rich & Co., A.G., was reeling in more than $200 million annually in profits.

With the overthrow of the Shah by the Ayatollah Khomeini, the major oil traders backed away from Iran, except for Marc Rich. Once the embargo was instituted by President Carter, all trade with Iran became illegal. But the Iranians, already deeply in debt, needed to sell, and Rich devised a scheme to funnel payments through his Swiss company. Placing him in an even more precarious position, Rich paid for much of the oil with weapons that he shipped in from Spain.

Rich and Green then got involved in the U.S. domestic oil market. They began selling discounted foreign oil to the major American producers, peddling it as their own. They hooked up with David Ratliff and John Troland, two Texas oilmen who were already making millions with this "daisy-chaining." When Troland and Ratliff were indicted and convicted, they fingered Marc Rich.

The government offered immunity to Rich traders willing to testify against their employer, and soon had a detailed understanding of Rich's operation. Rich had reportedly profited more than $70 million from this illegal resale of oil. But when called in front of a grand jury in early 1982, Rich refused to turn over crucial documents. He argued that as the top officer of a Swiss company, he did not have to comply with a court order from the United States. With scores of lawyers, Rich's team had the Justice Department tied in knots for months. Eventually he was found to be in contempt of court and ordered to turn over all documents or pay $50,000 per business day until they were received.

But on June 29, 1983, Rich dropped a bombshell. He sold his American entity, Marc Rich, International, to a new company, Clarendon, Limited. Clarendon was located at the same address and listed the same traders and staff, but Rich and Green were no longer on the board of directors. Clarendon would be headed by Rich confidant Willy Strothotte. A federal judge described this action as a "ploy to frustrate the implementation of the court's order."

By early August, Rich's attorneys had reached an agreement with the Justice Department on the contempt charges, paying more than $1 million in fines and agreeing to produce all documents. Yet only four days later the government received a tip that Rich was smuggling two unmarked steamer trunks of documents out of the country via Swiss Air. The government responded by seizing the documents and issuing a fifty-one-count indictment including charges of racketeering, mail and wire fraud, tax evasion, and trading with the enemy. On September 19, 1983, warrants were issued for the arrest of both Green and Rich. Later the indictment was expanded to a total of sixty-five counts. Preparing for the worst, in 1982 Rich had acquired Spanish and Israeli citizenship and applied to the Swiss government for citizenship. Pincus Green also acquired Israeli citizenship.

Meanwhile, Rich and Green were doing business as usual from their Zug offices, knowing that the Swiss government would not extradite them for tax evasion. Not only did the Swiss have a long tradition of providing a safe haven for financiers such as Rich, but Rich had also worked hard in Zug to cultivate a very different image of himself. As described in *Regardie's* magazine,

> After Rich fled his fourteen-room, $10 million Park Avenue apartment, there was a growing suspicion within the commodities community that his firm was also on the run, that it was running scared in the wake of mass layoffs because of what former employees might say when they were subpoenaed to explain the business. But the image soon changed. Rich, eager to present a kindlier and gentler face to his new neighbors, showered the Swiss with gold and goodwill. He spread $150 million among a foundation that teaches the disabled to work, a group that uses trained dogs to find people who are trapped in earthquake rubble, and a chamber

orchestra in Zurich. He also supports Zug's hockey team, a local high school's athletic complex, and various Swiss artists and photographers. He's given millions to Swiss museums. In 1987 he founded the Doron Prize Society, which awards a grant of five million Swiss francs annually to someone who's involved in 'charitable, humanitarian, cultural and scientific affairs.'

Rich's reputation for charity extended well beyond Switzerland. There was even a fellowship at Oxford University established in his name and a Marc Rich Chair in Economics and Business at the Universidad Carlos III in Spain.

The Clarendon company that Rich controlled was the same Clarendon mentioned in the Price Waterhouse Strategic Plan. Clearly, the lockout at Ravenswood was about more than Emmett Boyle and his petty resentments against the union. Compared to Rich, Boyle looked like little more than a schoolyard bully.

Rumors of Rich's involvement in RAC had been spreading in Steelworker circles for some time. Bowen and Becker knew of it, and Gary Hubbard asked Uehlein and Yeselson to look into the Rich connection. But the first substantive connection of Rich with RAC was actually made some months earlier by Linda McCoy of the Women's Support Group.

> For some time we had been aware that Marc Rich had been involved, and I took it upon myself to really find out who is this Marc Rich. There was one little piece of paper that circulated in the hall early in the year, and I began to go to the local library and check current article indexes to see if he's mentioned in any articles anywhere. And my husband and a neighbor and I went up to Charleston to the archives and tried to find out anything about Rich, and began to get copies of things about him, and couldn't believe some of it was in control of our lives here. We just couldn't believe it. And the people that worked at the company had no idea. Because, since my husband has been back to work, Ron Thompson, who is the department superintendent, commented to my husband in passing through his work area . . . that your wife probably knew more about Marc Rich than any of us here.

Among other things, Linda discovered a *Forbes* magazine article from April 1990 that suggested that, in the aftermath of Maxxam's leveraged buyout of Kaiser in 1988, "Rich's companies paid Hurwitz's company $435 million for three-year contracts on eight hundred thousand tons of alumina per year, plus a smelter and rolling mill in Ravenswood, WV."

Although Maxxam's general counsel sent a letter to *Forbes* denying the transaction between Rich and Hurwitz, even if Rich was not an official owner, clearly, through the Clarendon connection, he was still in control of the fate of the Ravenswood plant.

Once Uehlein and Yeselson had reviewed Price Waterhouse and the other documents, they contacted On Line Resources, a Washington-based database firm, to gather information on the original RAC buyout, the three major owners, and the corporate relationships. They also decided to do some field research in Stamford, where both Bradley and Strothotte's companies were based. In mid-February Rich Yeselson, along with Howard Scott from the USWA organizing department, traveled to Connecticut to glean information about Clarendon, Limited (Strothotte) and Stanwich Partners (Bradley) from county and district courthouse documents.

Bradley was an easy study. He had been involved in four breach-of-contract suits. He had had four liens placed on his $1.5 million home in nine years. And he owed almost a million dollars to the IRS. Yeselson suggests, "In other words, he was just a typical Wall Street sleaze bag. . . . This is the kind of stuff you want to find on your corporate targets."

At the same time, the local was gathering information on production at the plant. Charlie McDowell recalls, "Through the salaried people, we got all their customers and their shipments and their orders. I was given that on a weekly basis, the number of scabs that were in the plant, and the customers, and the pounds that were shipped, and what was returned." McDowell had been forwarding this information to Becker and the strategy group reviewed the material.

The campaign was beginning to take shape. March 12, 1991, marked the beginning of a new phase in the Ravenswood lockout. Until then, Chapman had been the International's spokesman to the local press, but now vice-president George Becker appeared for the first time, announcing a new initiative against RAC. For the previous four months the union had been largely reactive, placing its fate in the hands of the NLRB. For the first time, the union, under Becker's direction, was taking the offensive.

The campaign was twofold, targeting both owners and customers. At the corporate level, the strategy was to put pressure on absentee owners Bradley and Strothotte. Based on the information that Yeselson had gathered, the union sent a delegation of workers to Connecticut and New York City. A caravan of eleven cars containing forty locked-out workers arrived midday in New York to make their case directly to Bradley and Strothotte. They began by handing out a flyer to thousands of passers-by near the Clarendon office in Rockefeller Center. The head-line read "Stanwich Partners and Clarendon, Ltd. of Stamford, CT Are Destroying Our Livelihood in West Virginia at Ravenswood Aluminum Company—Why Do Charles Bradley and Willy Strothotte Want to Tear Our Families and Community Apart?"

Bill Hendricks was one of a small group of Ravenswood workers who went up to the Clarendon office, which he described as "an office which seems to be nothing more than a mail drop." Even though it was midday and midweek, the suite was dark and empty. Joe Uehlein, who had joined the group fresh from the *Daily News* campaign, which was wind-ing down, noticed that one of the names on the door, "I. Levantan," seemed familiar. He remembered later that night that in the book *The Metal Men,* which he had just finished, Ida Levantan was Marc Rich's personal secretary.

That night, the locked-out workers, several of whom were away from West Virginia for the first time in their lives, joined the *Daily News* work-ers at their victory party. Up at five the next morning, they were dressed in their "camo" and at the Stamford train station by six. They handed out the Bradley/Strothotte flyer, but they also passed out a very dramatic leaflet with a hand squeezing blood out of a Budweiser can. "Do not buy scab aluminum products," the flyer read. Referring to the recent deaths in the plant, the flyer continued, "The aluminum in those cans is tainted with the blood of hardworking men who lost their lives on the job. This Bud Is Not For You."

Later that day, the locked-out workers moved on to strip malls, shop-ping centers, and other public places in Stamford to distribute literature. Part of the group split off to picket in front of the Clarendon building. Because Uehlein was still tied up with the *Daily News* negotiations, a union public relations firm had been hired to do some advance work and set up publicity for the event. The firm had searched to find the "real" lo-cation of Clarendon, but when Joe arrived the workers were picketing the Clarendon Insurance Company, not Clarendon, Limited. Clarendon Insurance had no connection with Rich, Strothotte, or Bradley. Uehlein hurriedly took down the picket line and moved to the BHF (Berliner Handels and Frankfurter) bank, which was nearby and was a Ravenswood creditor.

After two more days of leafleting across the New York area, the locked-out workers returned to Ravenswood exhausted. The battleground had expanded. The fight at RAC was no longer restricted to Jackson County, West Virginia.

11

THE CAMPAIGN WIDENS

On Sunday, March 24, another Justice Caravan left Ravenswood for a second trip to Stamford. Close to forty locked-out workers piled into eleven cars and headed north to keep the pressure on Strothotte and Bradley. Once again they leafleted train stations and Clarendon and Stanwich headquarters in Stamford and New York, and were interviewed by local newspapers. On this trip they also leafleted shopping centers near Bradley's home ("Our Jobs and Lives Are Threatened by Charles Bradley of Stanwich Partners") and they urged shoppers not to buy beverages in cans made of "scab aluminum." The leaflet used the same design as before—blood flowing out of a beer can being crushed by a large, threatening hand—but this time the target was Stroh's beer.

During March Becker had brought the strategy team together to turn up the volume on the campaign. These meetings included Jim Bowen and Joe Chapman from USWA District 23; Bernie Hostein, Dallas Ellswick, and Howard Scott from the organizing department; Rich Brean, Paul Whitehead, and Carl Frankel from the legal department; Mike Wright from health and safety; and Gary Hubbard from communications. In addition to Joe Uehlein, Rich Yeselson, and Jim Hougan, Becker also hired Frank Powers, a Washington-based public relations consultant, and enlisted the assistance of Mike Boggs of the International Chemical and Energy Workers Federation (ICEF), the worldwide grouping of unions to which the Steelworkers belonged. The local leaders remained in Ravenswood. In the Steelworkers tradition, strategic bargaining decisions remained an international prerogative.

Becker chose this team because "they were original thinkers. . . . They come up with ideas. They don't take your idea and just agree with it and go on that." They came up with "different ways to screw things up and to go after them, different avenues, and ways to look at things." The team's task was to use the growing information they were gathering to develop multiple strategies to pressure RAC on every front.

As Uehlein remembers, the trips to Connecticut were in many ways "feel-good" exercises, "tactically doable, could involve a lot of workers. But

I think there were other things more important." As the strategy committee assessed the incoming information, they realized that the key to victory
did not lie simply in continuing the pressure in Connecticut. Instead, more
energy was focused on developing an "end-users campaign." These actions
pressured RAC's customers to withdraw their business and asked consumers not to buy RAC-made products, primarily beverage cans.

To track major customers, the local union for several months had been
"following the metal." Bill Doyle assembled teams of drivers who kept
their cars packed, gassed up, and ready to leave at a moment's notice to
follow trucks leaving RAC. They might be gone a few hours or a few
days. "They'd get there ahead of the trucker and wait and watch it come
in. Take pictures of the truck leaving the picket station. Take a picture
of them going in the plant," recalls Doyle. Because of their longstanding
connections to the workers, the truck drivers often disclosed their desti-
nations, which meant that Doyle's teams could meet the trucks at the
customer's location rather than having to follow them over hundreds or
thousands of miles. There they photographed the trucks again, to establish who was buying RAC aluminum.

Becker assigned Steelworkers staffer Dallas Ellswick to the end-users
campaign. By mid-March he had developed a comprehensive list of RAC
customers. USWA President Lynn Williams and Becker sent letters to several dozen companies, encouraging them not to buy "scab" aluminum.
They ranged from beverage companies, such as Coca-Cola, Budweiser,
and Miller, to major auto companies, such as Chrysler and Ford, and a
host of smaller manufacturers. The letter concluded, "Of course, the union
has a right to truthfully inform the public that your products are made with
scab aluminum produced by Ravenswood Aluminum Company."

Stopping the purchase of metal would stop the flow of money into the
company. "As far as we're concerned, every dollar spent . . . is a dollar
that's being spent to break the union," Becker said. In the short run, the
union developed a two-pronged attack. First, they would use their
influence to pressure can manufacturers whose workers were also Steelworkers. This would include efforts both by USWA locals at can plants
and by top union officials toward the CEOs of Crown, Cork and Seal
and National Can, two major can manufacturers. Second, the union targeted several of the largest breweries that used cans made from
Ravenswood aluminum. The first targets, picked late in March, were
Budweiser and Stroh's.

USWA District 33 director Dave Foster, who represents a large district
based in St. Paul, Minnesota, was at the Steelworkers executive board
meeting where this strategy was first announced, and he jumped at the
chance to get involved. "George Becker brought up that we were going

to start focusing on trying to put pressure on some of the end-users of the aluminum. And then we understood that there were a number of brewery companies that were involved in using Ravenswood aluminum in their cans. And he mentioned Stroh's. And that Stroh's had breweries in a couple of areas. And I stuck my hand up and said, 'Well, we've got a big brewery in St. Paul and I know the unions there. And we have the can plants that are making the cans, so we represent those can workers.'"

Reflecting the growing intensity of the campaign, arrangements were quickly made to send a Justice Caravan of locked-out workers to St. Paul. Arriving on April 1, they were featured the next day at a rally in support of Senate Bill 55, a piece of federal legislation to prohibit permanent replacement workers in labor disputes. Held in the rotunda of the state capitol, the rally was attended by newly elected Senator Paul Wellstone, a strong labor supporter and one of the signers of the bill.

The next three days were a blitz of leafleting across the Twin Cities. "We distributed twenty or thirty thousand fliers at every major intersection in Minneapolis and St. Paul, in the downtown areas, blanketed the University of Minnesota, which has a very large campus, leafleted at major plants at the shifts going in," recalled Foster. "And we included the telephone numbers of the brewery on there and it had an immediate impact, which surprised me, I guess. Because having been involved in a lot of these pressure campaigns, they usually don't spark quite as instantaneous a response as this did. The numbers of people that called were in the hundreds, just completely shut down the brewery's phone service."

In addition to leafleting, the locked-out workers met with union members from across the Twin Cities. Foster describes the first contact between his members at National Can and the workers from Ravenswood.

It was not a big meeting at all. There were thirty-odd people who came, because they work these crazy shifts. . . . We were crammed down in a little room, a dark room in the bottom of the VFW. And people were just getting off twelve-hour shifts, and were dirty and tired and came anyway because they were worried about their jobs and were worried about these people. And there was something deeply emotional about the connection those guys from West Virginia made with our people. I don't remember any of the things they said or the questions, but there was a kind of electrical connection between them and these can workers. No way were they going to be separated in this struggle. That night anyway, the people in the can plant would have given their jobs up to win that fight.

Foster, who had been deeply involved with the strike by UFCW Local P-9 against Hormel in nearby Austin, knew the importance of building coalitions and getting the Ravenswood story out. In a memo to Becker, he suggested that the local train several teams of workers in public speaking so they could go on the road to tell their story. The international developed a members' guide to public speaking and a guide to handbilling and working with the media. The union also set up a national toll-free hotline. Designed for Local 5668 members, the press, and supporters, the message began, "This is the United Steelworkers of America Ravenswood Lockout Hotline. The Ravenswood caravan is on the march again, this time to Minnesota." The message went on to describe the activities in the St. Paul area and to detail the lack of progress in bargaining back home.

Through February and March, Uehlein, Yeselson, and Hougan continued their research. They had gleaned what they could from databases; their next step was to gather information at a local level. Earlier, they had sent Dallas Ellswick to the Jackson County Courthouse in Ripley. They knew from the Price Waterhouse report that several loans used in the buyout of the plant might be recorded in the local courthouse. But Ellswick came up empty-handed.

Yeselson, who had spent a great deal of time in local courthouses all over the country, thought something was wrong. In early April he went down to look for himself. A tightly-wound, fast-talking, intense intellectual, Richard had completed his course work for a Ph.D. in history from the University of Pennsylvania. Rich Brean, who developed enormous respect for Yeselson's work, describes his first impression: "Yeselson [was] sitting there with like four or five days' growth beard. He looks like he's got a Molotov cocktail right there in the chair. He looks like the counsel of the IWW or something, you know, or a bomb thrower."

So when Yeselson arrived at the Jackson County courthouse in Ripley, West Virginia, he obviously was not a local.

They knew I was from somewhere else . . . he said, "You sure don't sound like you're from around here." I said, "You're right. I'm from northern New Jersey." He said, "Yeselson, what kind of name is that?" I said, "Polish Jewish." It's true. I mean, what are you supposed to say? "I's from West Virginia? Oh, you don't know our family, we're up in Parkersburg, over the hill, yeah, way over the hill, waaaay over the hill. God's country."

But the courthouse staff was polite to Yeselson. Checking the court records first, he found a workers' compensation case but little more. "So that was disappointing," remembers Yeselson. "So then I went to real estate and there it was, this loan agreement." Over a hundred pages long, the loan agreement detailed the financial arrangement that led to the leveraged buyout of Ravenswood Aluminum. Some of the information would prove very important later, such as the name of the law firm that had brokered the deal and the address of Rinoman Investments, one of RAC's holding companies. Yeselson also found confirmation of information the union had previously come across in the Price Waterhouse document: that at the time of the original buyout, Boyle, Strothotte, and Bradley had arranged a $140 million line of revolving credit with NMB Postbank, an international investment bank headquartered in Amsterdam.

Several aspects of the loan agreement immediately drew Marc Rich closer into the loop. The most important had to do with the Ridgeway Company, which was listed in the Price Waterhouse document as owning 48 percent of RAC. According to Yeselson, "The thing that you could pick up immediately from the loan agreement was that . . . all of Ridgeway's documentation . . . sent to Ridgeway Commercial pertaining to the agreement, was sent care of Clarendon Limited, Stamford, Connecticut. It was the classic bingo. So, in other words, Ridgeway is just part of Clarendon. And Clarendon, we know, is part of Marc Rich."

While many of the pieces were still missing, the Steelworkers were getting closer to unraveling the complex ownership of RAC and RAC's connection to Rich. Based on his research, Yeselson was convinced that Bradley was not important. "When we looked at this ownership structure, it was clear that even though the company was at first a hundred percent owned by Bradley, he was just sort of a facilitator." Strothotte, on the other hand, clearly was a major player, but, as the strategy committee began to realize, Strothotte's power could only be understood through Marc Rich. With information from the Price Waterhouse report and the loan agreement from the Ripley courthouse, the Steelworkers developed the first of many versions of a flow chart of the ownership and control of RAC. Revised as information came in, and later printed in four colors, the chart provided a quick visual for the strategy team and later for the press.

As the Steelworkers' campaign escalated, RAC fought back hard. On April 10, the company filed a civil suit under the Racketeer Influenced Corrupt Organizations (RICO) Act, against the local union and forty-seven individuals. According to the charges, "USWA Local 5668 and the individual defendants have conducted their affairs through a pattern of racketeering activities including violence and harassment against the corporation and the employees of RAC's contractors."

Although RAC's charges addressed the local violence, they were also clearly directed at the Steelworkers' larger campaign. An article in the

Jackson Herald reported, "Worlledge talked about letters sent to customers of RAC, and George Becker's coordinated campaign to hurt the company which included passing out pamphlets in New York City. 'The members of Local 5668 are more interested in destroying this company than bargaining,' Worlledge told the press. 'Their actions are not encouraging for those looking to provide jobs in Jackson County and West Virginia.'"

Among those named in the charges were activists such as Charlie McDowell, Mike Bailes, Bill Doyle, and Boyle's former neighbor and friend, Marge Flanigan. But the suit also included a number of people who had played little or no role in the local or in the campaign, and local president Dan Stidham was not named.

With the exception of the pipe bombings, RAC's charges referred to the kinds of petty incidents that often occur in a labor dispute, especially one with stakes as high as they were at Ravenswood. But these incidents hardly constituted racketeering and criminal conspiracy as the company suit claimed. The RICO complaint was almost absurd. More than 150 pages long, it listed 726 individual acts. One allegation read, "On or about January 15, 1991, defendants, through their agents and in concert with their co-conspirators, caused Greg Bowman's vehicle to be struck by an object as it passed the picket site at the south Y." Another read, "On or about November 8, 1990, defendants, through their agents and in concert with co-conspirators, . . . threw mud onto cars of Ravenswood employees attempting to enter the main gate." And "On or about November 25, 1990, . . . a grease ball was thrown at the vehicle of Harley Marcutu, hitting the hood and windshield."

Bill Doyle told the press that these incidents were exactly the same sort of indignities that members of Local 5668 had endured since the lockout began. In fact, many of the hundreds of incidents listed in the charge could just as likely have involved scab and security guard assaults on union vehicles and property as they did union activities against the scabs. Stuart Israel, who represented the union in the RICO case, suggested that the company had hurt its own case by including so many allegations. "By doing so, in my mind, they exposed the poverty of the allegations, because 90 percent of the actions were committed by unknown individuals."

Congress originally designed the RICO Act to prosecute organized crime. The focus was not on individual illegal acts, which were already covered under existing criminal statutes, but on how these illegal actions were part of a larger pattern of organized racketeering activity. The Act reads, "It shall be unlawful for any person employed by or associated with any enterprise engaged in, or the activities of which affect, interstate or foreign commerce, to conduct or participate, directly or indirectly, in the conduct of such enterprise's affairs through a pattern of racketeering activity or collection of unlawful debt." But over the past decade RICO has been extended far beyond its original intent, having been used against pornographers and even the Catholic Church (in connection with a cover-up of sex scandals involving three prominent bishops). In 1994 the Supreme Court ruled unanimously, in a landmark decision, that the National Organization of Women could sue anti-abortion groups under RICO. By dropping the original requirement that RICO could apply only where financial interests were involved, the Court greatly expanded the scope of the law.

In the early 1990s RICO was used by employers against new tactics developed by the labor movement. Texas Air brought a RICO suit against the Air Line Pilots union and the Machinists, alleging that the unions engaged in mail fraud when they conducted a corporate campaign against Eastern Airlines and distributed literature through the mail. In a suit against the Drywallers in California, employers argued that the use of strike funds to post bail for strikers arrested during demonstrations constituted criminal activity aimed to support further criminal acts.

The RICO suit had a chilling effect on the local and many of those named. In part this stemmed from the statute's requirement that guilty parties pay triple damages plus all court costs. Marge Flanigan reports, "I called an attorney which was a friend and a former judge . . . and he didn't think it was funny. He thought, you're in big trouble because it's written up as a conspiracy." Bob Buck and Gerald Church, under indictment for throwing a pipe bomb, were included on the RICO list. If they were convicted, Marge, as a "co-conspirator," could be implicated.

Marge seriously considered divorcing her husband to protect the home they had worked so hard to build. "The word came from inside the plant from one of the supervisors, 'Well, I don't agree with what the company did now, because they're literally going to try to destroy the people financially. You know it's bad enough that you've lost your jobs, now they're going after [you] financially.'"

Becker and the strategy group continued to explore pressure points. Safety and health remained key. In many ways the conditions in the pot room and the five deaths were the central events that led to the lockout. And scabs inside the plant continued to be injured at alarming rates.

During the first three months of the lockout, the Occupational Safety and Health Administration (OSHA) penalized RAC for conditions that had existed before the dispute began. On December 14, 1990, based on an investigation of heat stress stemming in part from Jimmy Rider's death, OSHA fined RAC the maximum of $10,000 for excessive heat in the pot room. According to the ruling, "The employer did not furnish a place of employment which (was) free from recognized hazards that were . . . likely to cause death or serious harm to employees, in that employees were exposed to excessive heat stress hazards. . . . These conditions were aggravated by lack of monitoring, lack of adequate training, lack of established work-rest regimens, lack of adequate medical surveillance and medical testing, and incomplete record keeping."

On January 28, 1991, RAC was slapped with a $27,700 fine for the July 18, 1990, death of David Evans. OSHA cited RAC for "willfully failing" to train workers to shut off power before replacing fuses. Evans was seriously burned while replacing fuses and died as a result. RAC was also cited for failure to require gloves and face shields while working at electrical stations.

Serious safety violations continued to be reported in the plant during the lockout, exacerbated by the large number of inexperienced workers in a dangerous industry. More seasoned workers knew about potential hazards and could react quickly when a few seconds meant the difference between life and death. Bill Doyle, chair of the safety committee, who had spent much of his life trying to make the Ravenswood plant a safer place to work, was particularly upset about the lack of training for the replacement workers and the company's record-keeping on injuries, now that the union was no longer in the loop. "I am sure the Ravenswood Aluminum Corporation Safety Group is continuing to alter safety records and statistics and not recording injuries properly as they were cited by OSHA in 1990 for record violations. The local union is aware of many serious injuries that have occurred inside the Works," Doyle told reporters. He went on, "One of my major concerns is that a major catastrophe in the Works could occur at any time with these unskilled, unqualified scabs. God forbid if my fears are materialized."

By March, USWA Health and Safety Director Mike Wright had pulled together a team of experts from the union. Building on Bill Doyle's work, they developed a strategic plan that included research, litigation, and actions around health and safety issues in the plant. To begin, Wright assigned staffer Jim Valenti to work with Doyle to explore filing formal complaints with OSHA. According to Valenti, the job was to

ferret out all the problems that [they] may have and to make them as visible as possible . . . I asked Bill if he could get a couple of people from the committee to just sit down with me and discuss what they did there. I got a big piece of paper and we kind of drew the outside of the plant, and I said, now what goes on in each of these buildings? And we talked about the fab side and the reduction side and the pot room and he tried to explain to me exactly what went on and what kind of equipment was in each of the buildings and what the process was from a chemical perspective, and that's how we kind of started, and I came back and I started doing some research.

Valenti also reviewed the health and safety records of the local, looking not only for individual incidents but for how they fit together into patterns of repeated violations.

After weeks of work, on April 26 the union announced that it was filing a comprehensive safety and health complaint with OSHA including pot room issues, the use of overhead cranes, and access to records. Although Local 5668 was still the legal bargaining agent for workers at Ravenswood, officials were denied access to health and safety information. Dan Stidham told the press that the replacement workers "have no protection whatsoever from management demands on them, no matter how unsafe . . . but somebody should protect them."

The same day, RAC responded by filing a defamation suit against Local 5668 and Bill Doyle, asking for a minimum of $10,000 in fines. The suit referred to the statements Doyle had made to the press in January about the company altering its health and safety records. Don Worlledge declared, "We consistently log safety statistics and injury reports, and I personally review all lost-time injury reports."

To dramatize the health and safety issues, the Steelworkers held a gathering of more than seven thousand supporters for a "Workers' Memorial Day" rally on April 28, 1991, at the Jackson County fairgrounds. Supporters came from as far away as Massachusetts, Missouri, and Illinois. This national day of mourning, established three years earlier by the AFL-CIO, saw rallies held across the country to memorialize workers who had lost their lives or been severely injured on the job. With the recent deaths in the plant, the event in Ravenswood was poignant. *The Jackson Star News* reported:

Lou Albright of the West Virginia AFL-CIO . . . read a list of "union brothers and sisters that had made the ultimate sacrifice as workers," losing their lives on the job. As Albright read each lost worker's name and West Virginia hometown, Charlie Cavender followed each name by striking a huge bell with a hammer, one resonating ring for each of the fallen workers. Each ring echoed across the crowded hillside and off the surrounding hills of Jackson County in a last tribute to twenty-five workers who died in West Virginia from April 1, 1990 to April 1, 1991.

Governor Gaston Caperton and USWA president Lynn Williams spoke, as well as Richard Trumka, president of the Mine Workers. Trumka reminded the crowd that Workers' Memorial Day "isn't about raising a little awareness, but it's about raising some hell, and we intend to raise some about what's going on in this country with safety in the workplace." He continued, "We're here today so we don't have to swap our paychecks for our lives."

But in many ways the focus of the day was best captured by Bob Rider, father of Jimmy. Rider told the crowd, "If Jimmy were alive today, he would be proud to stand here. Hopefully, there will be something that will come out of this to prevent a recurrence."

12

Small Victories

After what seemed like an endless winter, the Steelworkers' campaign picked up steam in the spring of 1991. As the weather heated up, so did the battle in Ravenswood. The Workers' Memorial Day rally was barely over when Emmett Boyle announced on May 10 that he was buying out his partners, Charles Bradley and Willy Strothotte. Although the announcement came as a complete surprise, company officials went out of their way to insist that this decision was unrelated to the lockout. "This was something discussed long before the strike at Ravenswood," Debbie Boger told the press. "It should have no impact on the Ravenswood situation at all."

Industry analysts, however, were not convinced. "They've reshuffled the cards, but who's holding what is not clear," analyst Peter Merner said. "You never knew for sure who was financing the original deal; you don't know for sure who's financing this one." Given Marc Rich's long reach in the aluminum industry, his involvement in the Ravenswood buyout was far from clear. But a buyout by Boyle represented a significant risk to the locked-out workers. Beginning with his defeat in the pot room in the early 1980s, workers had watched Boyle's intransigence grow. He had pushed through the job combinations and cost-cutting that precipitated the lockout. As the stand-off dragged on, Boyle seemed to take events more personally each month, from the graffiti painted on barns and roads to the drive-bys by the Women's Support Group at the plant gate. At the plant he was growing increasingly impatient and seemed to be willing to risk everything to break the union and win.

Marc Rich, on the other hand, was ruthless and powerful, but he had no personal stake in Ravenswood. More important, he was a consummate deal maker who depended on operating in secrecy. Despite his power, this need for secrecy made Rich vulnerable to outside pressure. If Boyle succeeded in buying out Strothotte, Rich's connection to Ravenswood would be severed and this important leverage point lost.

The first break in the NLRB case came in mid-May when NLRB officials determined that RAC had bargained in bad faith by failing to provide the union with the results of the heat study on the pot rooms, despite Joe Chapman's repeated requests. This was illegal because under NLRB standards, the union had the right to information relating to health and safety, a mandatory subject of bargaining. The following week the NLRB ruled against RAC again, this time charging that the guards had taunted union members, assaulted them, and harassed them by closely following vehicles and by unlawful surveillance of the union hall. Given the members' absolute hatred of those Judge Fox had referred to as "arrogant black-suited military strangers," the ruling against RAC's use of the goon guards was extremely important to them. The Steel-workers hoped the NLRB rulings would force RAC back to the bargaining table. As Jim Bowen put it, "This has got to end sometime and now is as good a time as any."

With RAC on the defensive, OSHA officials arrived the next week, on May 23, to inspect the plant. Bill Doyle's refusal to let go of safety and health, even if his members were no longer in the plant, had finally paid off. The charges that Jim Valenti had filed, coupled with the earlier deaths and fines, convinced OSHA that there were serious violations. But when the inspectors arrived at the plant they were denied access, despite their warrant. RAC management offered their justification in a terse press release: "Since this complaint appears to have been filed by people who have no current knowledge of conditions in the plant, it is unwarranted and has no probable cause. Under these circumstances the visit by OSHA appears to be for the purpose of harassment. RAC will not be a party to the manipulation of a government agency."

RAC's indignation, however, was unfounded and ill-advised. As far as OSHA was concerned, the USWA continued to be the authorized union representative for RAC workers, regardless of whether they were on strike, locked out, or nonmembers working in the plant. As area OSHA director Stan Elliott explained on his way to discuss the case with a federal magistrate, "The employee representative in this case, the certified bargaining unit, is the Steelworkers. The Steelworkers have a right to protect its membership and protect employees that are in the plant. We will go to court to enforce the warrant."

The union response was swift. "Our reaction is outrage," proclaimed Jim Valenti. "Denying OSHA entry when they have a warrant is not common at all, especially when the company is trying to portray the image they're totally safe."

District director Jim Bowen went further. "If any corporation in America can high-handedly deny entry to officials of the federal Occupational Safety and Health Administration who have a legal warrant requiring that access be granted, then every American worker's right to a safe job is in jeopardy." Ironically, less than two hours before the OSHA inspectors were turned away, a replacement worker, James Seaman, was seriously injured when his chest was crushed while he was moving a piece of machinery.

RAC's refusal to admit OSHA inspectors allowed the union to seize the moral high ground for the first time since the early days of the lockout. While they knew that RAC had acted wrongly and callously during bargaining, that process was mostly hidden from the public eye. The goon guards and the petty violence had been turned back on the union. But here was a simple, very public example of RAC, in the face of continuing accidents, turning away the federal agency that had the legal right to inspect the plant.

The situation created not only favorable press but also an opportunity to bring political pressure on RAC. The union began calling on politicians at all levels. Although many public officials had expressed support for the locked-out workers, given current labor law there had been little opportunity to offer more than words of sympathy and outrage. But RAC's flagrant violation of the law gave a number of politicians an opportunity to show their support. The *Parkersburg News* reported, "Wearing 'Fort RAC' shirts and, in some cases, jackrock earrings, about a half-dozen wives approached [U.S. Senator Jay] Rockefeller upon his arrival at a senior citizens conference in Parkersburg. The senator seemed surprised when the women handed him a newspaper article describing the company's decision last week to deny admittance to two Occupational Safety and Health Administration inspectors... 'I find that incredible,' Rockefeller later told reporters when asked about the incident. 'That I am going to check into.'"

The next day, Representative Joseph M. Gaydos (D-Pa.), chairman of the Subcommittee on Health and Safety of the House Committees on Education and Labor, announced an investigation of RAC's safety and health record. The committee, said Gaydos, was aware of only a handful of instances where management had refused entrance to OSHA. Meanwhile, Jim Bowen was pushing for a complete "wall-to-wall" inspection of the plant, a procedure used by OSHA only in extreme cases.

On the defensive, RAC took out a full-page ad in local newspapers. The headline read: "To those who want to know the other side of the story." The ad argued that typically in labor disputes employers are seen as sprawling giants while unions are the "folks next door." Despite the fact that RAC was part of a major aluminum conglomerate through ORALCO and its connection to Clarendon and Marc Rich, the ad contended that the roles were reversed in this case. "The USWA, a giant organization with hundreds of thousands of members, is using all of its resources against Ravenswood Aluminum Corporation." Intimating a conspiracy by the union, the ad concluded, "Is it possible that those who are trying so hard to destroy Ravenswood Aluminum Corporation really want to come back to work for this company? Or has this gone far beyond the seventeen hundred USWA members on strike?"

Meanwhile, the end-users campaign also gained momentum. On May 4, a twenty-member Justice Caravan IV traveled to Louisville to handbill the Kentucky Derby. Later in the month, fifty locked-out workers leafleted the Indianapolis 500. Working with staff and members from District 30, they distributed thousands of the "bloody can" leaflets urging road racing fans to tell Stroh's, Old Milwaukee, 7-Up, RC Cola, and Diet Rite Cola to stop using scab cans.

From the beginning, the end-users campaign entailed a risk. The Steelworkers could legally inform can manufacturers about the events at Ravenswood and ask those companies not to buy RAC metal. But since enactment of the Taft-Hartley Act in 1947, it has been unlawful for a union to call a boycott of a company simply because it does business with an employer where a labor dispute is in progress. Consequently the Steelworkers could not legally ask consumers to boycott beverage companies because they were using Ravenswood aluminum in some of their cans. They had to be certain that the cans were made with Ravenswood can stock. "In that area, we fought the lawyers," recalls George Becker. "This was at the cutting edge of as to whether it may be legal or illegal. . . . Rich Brean is our specialist in libel law. I love him, but he used to drive us crazy. I mean, some things are not without risk and we couldn't do the things we wanted to do without taking the risk, so we were willing to take the risk. Rich helped us minimize the risk. But in the end, we had to do it if we were going to win."

Given these high-stake risks, Becker wanted to make sure that the information he was receiving was ironclad. Dallas Ellswick, the staffer on the end-users campaign, recounted, "George said to me, 'Dallas . . . we're more than likely going to be sued, but I'd rather get sued for something you say took place than for something we heard third or fourth hand.'" At first Ellswick was just sorting out the data passed on to him by the local's truck tracking team, but as the end-users campaign heated up that spring, Becker ordered him to take direct charge of the information gathering and the truck tracking.

The assignment created problems with the local union. In addition to his health and safety work, Bill Doyle had spent months with a dedicated core of volunteers working tirelessly, tracking virtually every truck that left the loading dock. Charlie McDowell had been passing along inside information on where metal was being shipped. Charlie and Bill were certain that their information was correct. In cases where they did not know where the trucks were going, a team would try to follow them all the way to the end. But more often than not they knew the truck's destination from information leaked by sympathetic salaried workers inside the plant or from the truckers themselves. In those cases, they would simply photograph the truck, including the license plate number, as it left the gate, and then meet the driver at his destination and photograph the truck and license plate again.

But now Becker and Ellswick wanted information that was foolproof. They wanted the trucks followed every mile to their destination and they wanted photographs, not just of the truck leaving the plant and reaching its destination, but, also, whenever possible, of the metal being unloaded from the truck, with the RAC label clearly visible. Ellswick explains, "What we wanted to do was to verify beyond the shadow of a doubt that the metal was transferred from the Ravenswood gate and it went to this place here." He told the workers, "If you don't see it you don't sign. . . . I want to see your name saying that you saw this truck go from this plant to that plant and seen it unloaded. Because if you don't see it, and we get sued . . . we don't want some lawyer saying, 'But you really didn't follow the truck, did you? You don't know that that's the metal that came from the plant?'"

Becker and Ellswick decided that they didn't need to follow every truck; instead the resources should be dedicated to gathering rock-solid information on the major can and beverage companies that were the targets of the end-users campaign. Once they had solid information on one company, there was no need to continue tracking trucks delivering to that company. This policy also met resistance from the local. "[A] lot of them wanted to go out and follow every truck," Ellswick explains. "They felt that if we didn't follow every truck then we weren't doing much good. And they really didn't understand that the way we were doing it, we weren't trying to stop everything. What we were trying to do was hurt them [RAC] moneywise. And you do that by stopping 30 or 40 percent from now on. And you say you're not going to stop."

Although tempers continued to be hot in the local because of Ellswick's intervention, he assigned Clinton Durst, chair of the workers' compensation committee, to assemble teams of drivers for this new phase of truck tracking. A quiet, soft-spoken man, Durst helped Bill Doyle hold down the fort, day and night, whenever Stidham and McDowell were in meetings, in negotiations, or on the road. Durst put together a list of fifty to sixty people that he could call on very short notice to grab their bags and hit the road, sometimes following the trucks for days on end. Durst staffed the phones at the union hall, waiting for his teams to report back on where they were and what they had found. Under Ellswick's direction, Durst hounded the trackers to keep accurate records and do everything possible to stick with the trucks and get clear photographs of the metal being unloaded at the destination.

Many of the truckers came from the Ravenswood area and still others supported the locked-out workers' cause. Gene Fowler describes spotting a trucker they were looking for in his cab at a truck stop.

The boy was asleep. So we went back to the little restaurant and I got him a big cup of coffee. We waited about a half hour probably and I tapped on the window. He was in his sleeper, he stuck his head out there and I said, "You ready for some coffee?" So he started drinking that coffee. He said, "I got in here at three o'clock this morning and they wanted to unload me at three o'clock. If I'd a let them, you wouldn't've got no pictures, so I told them I was going to sleep a while before I unloaded." See, so we got pictures there. We got pictures there of him unloading there in Atlanta.

Meanwhile, Doyle and his smaller team continued to track the metal their own way, convinced that following the trucks port to port was not necessary, especially since only one driver had ever failed to end up at the expected destination. But as Doyle tells it, "I never did say, 'Hey, I'm not doing it your way.' All I did was say 'Okay.'" He simply continued taking the trips, taking the photographs, and passing the evidence on to Ellswick and Becker.

Becker and the strategy group were beginning to use the information to put the squeeze on several large can manufacturers and beverage companies. Becker describes,

We targeted Budweiser, and Budweiser told us they don't use Ravenswood Aluminum, and that if we went forward with this, they were going to sue us. Our lawyers got extremely concerned when they said that. They said,

The Ravenswood Aluminum plant.
Photo: Michael A. Lucas.

Jimmy Rider gravestone.
Photo: Dale Ferrell.

March of the goon guards, November 25, 1990.
Photo: USWA Local 5668.

Graffiti.
Photo: Dan Stidham.

"Rally in the Valley," December 30, 1990, Charleston, West Virginia.
Photo: Dale Ferrell

One day longer.
Photo: Frank Powers, FP&S.

Local 5668 president Dan Stidham, Workers' Memorial Day, Jackson County
Fairgrounds, April 28, 1991.
Photo: Dale Ferrell.

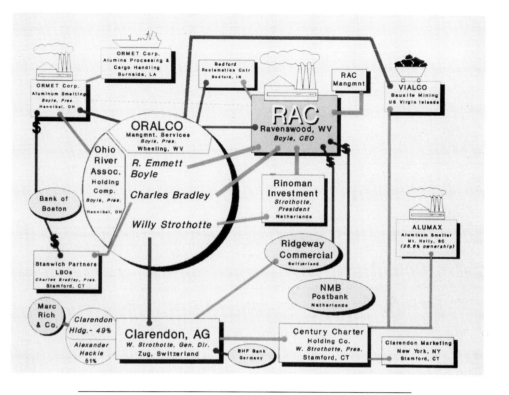

Ravenswood Aluminum Co.—A hidden chain of international ownership.
Graphic: USWA.

USWA president Lynn Williams, Workers' Memorial Day, April 28, 1991.
Photo: Dale Ferrell.

DON'T BUY:
Stroh's, Old Milwaukee in cans made with scab aluminum.

Death stalks the workers at Ravenswood Aluminum Corporation (RAC)— a company that produces aluminum used in some beer cans.

Five workers were killed at RAC's West Virginia plant in the 18 months after it's new owners took control. Under the previous owners, only 1 work-related death occurred in 20 years.

When it came time to negotiate a new contract, RAC wanted the 1,700 members of the United Steelworkers of America to trade life and limb for a paycheck.

The workers knew they had a right to a safe job with decent pay and benefits. But the laws protecting American working people apparently mean nothing to RAC.

On November 1, 1990, RAC locked the workers out of their jobs. Replaced them with scabs. Turned the plant into an armed camp surrounded by barbed wire, guarded by RAC's private army.

This is an attack on every working American

Corporate America is watching what happens at Ravenswood. If RAC can get away with lax safety rules and stealing jobs, other companies can, too.

And then no American job will be safe... *including yours.*

You can help the proud workers at Ravenswood and their families in the fight for American jobs. Tell the beer companies: **stop using bloody, scab aluminum.**

Before you buy, ask if your can is made with scab aluminum.

For more information, contact:
The United Steelworkers of America
Ravenswood Coordinated Campaign
P.O. Box 56 • Ravenswood, WV 26164 • (304) 273-9319

Flyer distributed as part of end-users campaign.
Graphic: USWA.

Handbilling at Churchill Downs, May 4, 1991. Bill Hendricks, left.
Photo: USWA.

Women's Support Group welcomes one of the many union caravans to Fort Unity.
Photo: Charles Robideau / USWA.

District 23 director Jim Bowen rallies the crowd, Fort Unity.
Photo: Charles Robideau / USWA.

One of the many meals served at Fort Unity.
Photo: Gary Hubbard / USWA.

WANTED

MARC RICH

✦ FUGITIVE ✦

$750,000 REWARD

Marc Rich, head of the Marc Rich Group Companies, is wanted by the U.S. Government, which is offering a $750,000 award for his arrest, on a 65-count indictment for:

- ✦ Tax fraud
- ✦ Racketeering
- ✦ Conspiracy
- ✦ Trading with the enemy

Rich is also alleged to have:

- ✦ Traded grain to the Soviet Union during the US grain embargo of 1979-80
- ✦ Shipped oil to South Africa during the international trade embargo

HIDEOUT: Zug, Switzerland

WARNING: Apprehension of suspect may be EXTREMELY DANGEROUS—alleged to be protected by mercenaries armed with sub-machine guns.

Marc Rich "Wanted" poster.
Graphic: USWA.

"Solidarity with the American Aluminum Workers." Dewey Taylor, Dan Stidham, Joe Chapman, Joe Uehlein, Beat Kappeler (secretary of the Swiss metalworkers union), Charlie McDowell, Mike Bailes, Joe Lang, and Jack Wheeler in front of Marc Rich headquarters, Zug, Switzerland, June 1991.
Photo: USWA.

Charlie McDowell makes the Clarendon connection. Marc Rich headquarters, Zug, June 1991.
Photo: USWA.

Mother Jones chastising Marc Rich in Rotterdam, March 1992.
Photo: Dewey Taylor.

FRATIA rally in Romania, February 14, 1992.
Photo: Penny Schantz.

Randall Frye (NLRB Region 8 director, left), George Becker (center), and Pete Nash (right) sign the final contract and the settlement of the NLRB charges, June 1992. Photo: Gary Hubbard / USWA.

Returning to work, June 29, 1992. From left, Joe Uehlein, Joe Chapman, Joe Powell, George Becker (with raised fist), Dan Stidham, and Jim Bowen. Photo: John Himelrick.

George Becker.
Photo: USWA.

Marge Flanigan, return to
work, June 29, 1992.
Photo: Bryce Turner / USWA.

Mike Bailes, back through the gates, June 29, 1992.
Photo: Gary Hubbard / USWA.

"Don't do it." But our decision was that we're going to do it, because we knew we were right. . . . Now, at Budweiser, this guy, it turned out, was from my home town. I used to work for his mother. She run the local sandwich shop. . . . We had some pleasant discussions. And he told me, "I'm telling you, George, there is no Ravenswood aluminum being used at Budweiser. And our legal department is going to go right down your throat. We'll break you. Please, don't do this." And I said, "Well, I'm going to tell you something. We're going to do it. It's coming out this afternoon, but first I would like you to call this plant down South and talk to this supervisor. Don't let them put you on to anybody else. You talk to him. Then if you still feel that Budweiser's not using RAC aluminum, call me back and I'll pull the thing." And he called me back. He said, "George, I want to make a deal."

The hard work was beginning to pay off. Nobody had caved yet, but the Steelworkers were very close to getting a number of companies like Budweiser to stop purchasing RAC aluminum.

As the months dragged on, Local 5668 members' finances were drained. While the Assistance Center took care of mortgage and utility bills, savings dwindled and personal budgets grew tight. The Assistance Center started picking up the slack and distributing food on a larger scale. The volunteers started slowly, with a few shipments from Mountaineer Food Bank and a few donations of bulk food. "We had fifty-pound bags of Granola," Bud Rose recalled. "They would make arrangements with the Women's Support Group and they'd send down eight or ten women. They'd go down there and bag it into five- or ten-pound bags. . . . Hartley Oil brought in two boxcars loaded with potatoes that had been rejected somewhere out in the Midwest. I don't know how many pickup truck loads of that we had."

Flowers Bakery, owned by the in-laws of pot room worker Mike Schmidt, donated large quantities of bread and baked goods throughout the lockout. Every day, locked-out worker Don Lipscomb and his wife Leanna picked up bread at Flowers and delivered it to Ravenswood. This became a full-time volunteer effort. Don describes the daily routine: "She would call us and she knowed that when she called us we would be there within the hour. . . . I just got a Chevrolet pickup with a cover on it. And I put over a hundred trays in there. . . . When we got it to the hall, people would help unload it. We didn't have no problem getting it unloaded. She would give potato chips and cakes and pies and just anything that a thrift bakery store would have."

The Lipscombs were not the only people who volunteered to take on the mundane chores that were a necessary part of keeping the workers together and the union functioning during the lockout. With the heavy traffic in the union hall, one of the members came in every morning and mopped the floors, cleaned the bathroom, and emptied the trash. Several women workers helped out in the office to process the faxes and phone calls from across the country that flooded in and out of the union hall. Other locked-out workers spent hours in a special radio room set up in a back room of the hall, listening to and painstakingly recording every radio message they could intercept out of RAC.

Under Doyle's and McDowell's direction the union also monitored the plant from the water and from the air. As Doyle describes the operation, "We had our own planes, our own air force. We had our guys in the water, our own navy. . . . Any time we wanted to see what was going on, we'd just jump in the plane. . . . We knew what was going on in the river. We had boats." Locked-out workers posing as fishermen would spend hours floating up and down the river, watching the plant and reporting anything unusual back to the hall.

But other than regular picket duty, the main volunteer work was at the Assistance Center. As the local began to receive more bulk and surplus food, the Assistance Center had to figure out a way to store and keep it. "We started getting a lot of donations from rallies and stuff. Different locals would be coming in here with trucks of food," Glen Varney explains. "I told Dan, 'This is too much, I can't do that.' I said, 'We need another person.'" Locked-out worker Joe Strickland was assigned to the task and, like Rose and Varney, he was an excellent organizer. First, drawing from their own members who had been skilled workers in the plant, they poured a concrete floor. Strickland remembers, "They got the floor laid and I said, 'I've got to have some shelves in here.' Well, Larry Mott, he come up and said, 'Well, I got lumber enough for this.' Then food started coming in and they had this big lot of tuna, and we didn't have enough space to put it. I got hold of Wayne Fletcher. I said, 'We need to have a loft up there.' He said, 'Yeah, we'll get it up there.' The next day he was down there, you know, putting the loft up."

Joe Chapman had told them they would be "begging, bumming, and borrowing" before they were through, and that's exactly what Strickland and his crew did. When they needed a refrigerator or a freezer, they put the word out, "and after a couple of days, somebody would say, 'I got one you can use.' Everything we had up there was donated. I got hold of the Coke company up there. . . . I said, 'We need a big cooler' . . . and just a couple days later they brought it down and set it up for us."

Once the basic work was done getting the Assistance Center ready, Strickland pulled together a crew of over a hundred to unload, stock, and distribute food. But the donated food did not always translate into meals, and the pressure was on to do more. Strickland approached Jerry Carpenter, manager of the local Foodland, which was next door to the Assistance Center. Carpenter agreed to purchase food in bulk for the Center.

I don't mainly want to get involved in labor disputes, but the fact of it was, it was not handled right from the beginning. People were illegally locked out of the plant, and you turn around and you look at who has built this community. Yes, Kaiser Aluminum came in and put a plant in the area, but it wasn't Kaiser Aluminum's money that built the town and the businesses that are here. These businesses are built by the people that work at that plant and that support these businesses. Now, if they get locked out and I turn my back on them, I don't do anything but cut my own throat.

Carpenter sold food to the Assistance Center at cost and even partitioned the back of his store to help with its inventory. The crew from Local 5668 not only unloaded their own food but often helped Carpenter as well. With this regular source of a variety of food, the Assistance Center became a full-fledged food bank. Meals were planned and members were assigned a weekly pickup. While not everyone in the local used it, the Assistance Center was crucial for many families to hold out until the lockout was settled.

With the national media attention they were receiving, and the many trips and speeches that Local 5668 members made to other unions, donations started pouring in. In addition to the strike benefits, these donations were critical in keeping the Assistance Center alive. "We wouldn't have been able to pay somebody five dollars' expense for driving up here to the picket line," says Bud Rose. "We wouldn't have been able to give them a twenty-dollar bill to buy their kids an ice cream cone, if it hadn't been for the support from the coal miners, the auto workers, the teachers associations, and just private donations. . . . We'd get maybe five dollars or ten dollars from somebody that read about us in California or Oregon or Maine."

The activities of the Women's Support Group had evolved as well. "After the drive-bys stopped, we were going to fall apart because we'd been so active and we were meeting once a week," worried Marge Flanigan. "We decided that we were not going to let the company think they had defeated us. No matter what they did, we always came back with another program." So once the weather broke, the Support Group began regular suppers outside on the grounds behind the union hall. They started with wiener roasts over open fires but, according to Marge, "We soon got sick of hot dogs. So we decided then we would go to covered dishes. So we set up tables and had covered dish dinners." By May these Tuesday night dinners had become an institution in Ravenswood.

But the accommodations for anything more than a picnic were pretty primitive. Buoyed by the enthusiasm generated on Tuesday nights, the union began construction of "Fort Unity." In answer to "Fort RAC," they started building on a flat piece of land on top of the hill behind the union hall. As Marge Flanigan tells it, "The union realized real quickly that we were serious, that we weren't going to quit, so they decided to build us a kitchen. And they said, 'Would you like a picnic shelter and a kitchen and a stage?' And we have Fort Unity."

An enthusiastic group of volunteers went to work, organized by Jerry Butcher, who took a break from tracking trucks. "He was just the type of person that, if you needed something done, Jerry was there," says Marge Flanigan. "He could organize it. He could get it done. And he never had a complaint. He was always friendly, always cheerful, always full of positive, positive action. . . . He could make things happen and he played a major role in building Fort Unity."

Joe Chapman recalls, "They jumped in and they knew how to do that. They took care of whatever you asked them to do. They were really innovative." By midsummer Fort Unity was completed and provided ample facilities for the Tuesday evening get-togethers, including a large rally on Memorial Day. "We may be locked out of our jobs, but we sure eat good," reported wife Libby Ashby.

One of the regular features of the Tuesday nights that summer was the Paw Paw band, fronted by negotiating team member Dewey Taylor.

> Joe Chapman knew that I'd done a little of it and he talked to me. "You know," he said, "we're gonna need something like this." And I said, "I'll get some people together." And I had another old fellow that I'd played a little with, that retired from the plant, who was a good bass player and lives just out of town here. Well, next time I came, why, here comes a guy down across the lot with a guitar, Paul Fitzpatrick. He gets up and he does a couple songs. And the next time some other guy brings his fiddle and he gets up and does a couple things.

The band was called the Paw Paws because all the members were grandfathers. They played a blend of bluegrass and country and even rewrote some old songs with new lyrics for the lockout.

The Assistance Center, the picnics at Fort Unity, the truck tracking, and the trips to Kentucky and Indianapolis certainly appeared to be working to hold the union together. Six months into the lockout, still only seventeen of the seventeen hundred workers had crossed the picket line.

In April, former local president Gene Richards had crossed over. Dan Stidham recalls, "He was up for president of his local union, and when he didn't get that, there was nothing left for him." Richards had run against Stidham and was voted down 736 to 39. "So he crossed the picket line and went back in the plant. His wife's working salary. His brother is assistant plant manager down there. He's got a brother that's a foreman. So he had a lot of people on that side of the fence. . . . The word is pretty straight that he had committed to bring as many as 250 back to work. He took one with him, and then two or three weeks later his brother went with him. So his brother scabbed and went back too."

Becker and the strategy team knew that, even with the picnics and the Assistance Center, others would soon be tempted to waver. Because of his organizing experience, Dallas Ellswick was drafted to develop a grassroots housecalling campaign to shore up membership support. By the end of April, Ellswick had put together a plan: "I set in motion a system to housecall [visit at home] everybody up there at that plant that wasn't actively involved in that work stoppage. I put everybody's name on the computer. And then I went to the picket captains and I had them to tell me everybody who had served picket in there. And I took that and deducted the ones that didn't. And I set up areas, like if it was Point Pleasant area, I put a code for Point Pleasant. And I punched the computer and run off everybody in Point Pleasant who hadn't been active in the work stoppage."

After running the numbers, Ellswick discovered that about two hundred workers, a little more than ten percent of the local, weren't actively participating. With the help of Dewey Taylor, he enlisted twelve two-person housecalling teams, each assigned an area. The teams went through a training program where they practiced going through a list of questions that Dallas had put together, and were given report forms to turn in after each completed call.

The work was painstaking, but all those on the list were visited. And the visits paid off. The callers found that most of the two hundred people on Dallas's list had perfectly good reasons for not showing up for picket duty. Some had found other jobs, several had retired or were on disability, others were devoting so much time to volunteering at the Assistance Center that they were not available to picket. If any had been thinking about crossing the line, the housecalls seemed to help. That spring Gene Richards and his brother were the last to scab. The rest of the seventeen hundred held firm.

13

TURNING TOWARD EUROPE

Throughout the spring of 1991, Joe Uehlein, Rich Yeselson, and Jim Hougan continued to collect and analyze information on Marc Rich and his empire. They slogged through overflowing printouts, reams of data, and piles of articles from the ICEF international database put together by Michael Boggs. They analyzed and reanalyzed the loan agreement, and read and reread the information Hougan had gleaned from individuals tied to Marc Rich. Piece by piece, they were beginning to put together the vast labyrinth that constituted Rich's empire in more than twenty countries in Europe, Asia, North and South America, and Africa. As the spring progressed they became more and more convinced that the campaign had to move to Europe to pressure Marc Rich.

From their research they discovered that Rich's interests went well beyond oil and aluminum to include copper, zinc, gold, lead, and agricultural commodities such as wheat, barley, corn, sugar, and cocoa. He was also invested heavily in the real estate and tourism industries in Spain and Eastern Europe, where he was known for purchasing properties on the verge of bankruptcy and reselling them for a multimillion-dollar profit.

"He is a titan in the business of wholesaling the planet's natural resources to the highest bidders," explains Jim Hougan. "He owns or controls oil wells in Russia, mines in Peru and electrical supplies in England. There are refineries in Romania, office buildings in Spain and smelters in Australia, Iran, Sardinia and West Virginia. He has forty offices and thirteen hundred employees throughout the world and is simultaneously the uncontested emperor of aluminum, a prince of sugar, a shogun of soy, a mover and shaker of the world's markets in nickel, lead, zinc, chrome, magnesium, copper and coal."

Despite having a controlling interest in almost every metal and agricultural commodity on the world market, there was very little that Marc Rich owned outright. From zinc mines to aluminum smelters to luxury hotels, Marc Rich & Co., A.G., owned between 20 percent and 49 percent of each company, but very rarely more than 50 percent. More often than not his partners, Willy Strothotte, Felix Posen, and Alexander

Hackel, had the controlling interest, rather than Marc Rich himself. It was to Hackel, for instance, that Rich transferred 51 percent of his American subsidiary, Marc Rich, International, which was renamed Clarendon after Rich was indicted by the Justice Department in 1983.

This innovative structure allowed Rich to rule his empire with only thirteen hundred employees, far fewer than trading firms a fraction of his company's size. It was an arrangement that made proving the connections between a company such as Ravenswood Aluminum and Marc Rich himself extremely difficult. The distance also provided Rich with excellent cover when deals went bad. Perhaps most important, this structure allowed him to establish profitable relationships with businesses and governments who would have been quite squeamish about a more visible association with one of the world's most notorious white-collar criminals.

Although he made his fortune in oil, and had become heavily involved in the world grain market, aluminum remained at the center of Rich's business activity, earning him the name "Aluminum Finger" in the metals trading community. In the early 1980s Rich had begun to buy a controlling interest in sources for both bauxite and finished aluminum. He concentrated his efforts on establishing lucrative relationships with countries such as Venezuela, Iran, and Romania, which were rapidly moving into the aluminum smelting arena. Despite his huge presence in the industry, he had direct ownership interest in only two aluminum plants: 27 percent of the Alumax smelter in Mount Holly, South Carolina, and 21 percent of the Eurallumina plant in Sardinia.

Rather than directly purchasing smelters, Rich, and Rich subsidiaries such as Clarendon, established "tolling agreements" through which they paid smelters to process alumina from Rich's own supply in the Virgin Islands and Jamaica, and then Rich and/or Rich subsidiaries would own and sell the finished aluminum. This was exactly the kind of arrangement between Clarendon and Ravenswood Aluminum that Chapman discovered when he was investigating the profit-sharing plan.

These arrangements were extremely profitable for Rich. At the Mount Holly plant alone, after an initial investment of just $40 million to purchase a one-quarter interest in the smelter, Rich, through a tolling agreement with Clarendon, sold the fifty thousand tons of aluminum produced each year at an estimated annual profit of more than $65 million.

By the early 1990s, Rich moved more than 2.5 million tons of aluminum a year, more than 40 percent of the six million tons of aluminum traded on the world market, according to industry estimates. As markets suddenly began to open in Russia and Eastern Europe, Rich stood poised to lead the "gold rush" to take over the metals markets in those nations. By the late 1980s, according to one senior trader from a competing Australian metals conglomerate, "In metals, it's now Marc Rich and the forty dwarfs."

Rich's bauxite operation in Jamaica was a key ingredient of many of these deals. In the early 1980s he had developed a close relationship with then Prime Minister Edward Seaga, loaning the Jamaican government more than $200 million and donating $45,000 to send Jamaica's track team to the 1984 Olympics. In 1985 Rich and Seaga signed a ten-year agreement whereby Rich would purchase more than four million metric tons of Jamaican alumina at less than half the market rate. Three years later Rich underwrote the cost of sending the Jamaican bobsled team to the 1988 Winter Olympics.

In 1989, Seaga lost his office to Michael Manley, a socialist legislator from the People's National Party. Educated at Oxford, Manley was the son of Sir Norman Manley, a former Jamaican prime minister himself. Despite this background, Michael Manley had begun his career as an organizer with the Bauxite Workers Union. There he had developed a lasting relationship with the Steelworkers, who had aided the Bauxite Workers in their organizing drive. A central theme in Manley's campaign was Rich's close ties to the South African apartheid regime. Just as Rich and Pinky Green had brazenly violated the oil embargo against Iran a decade earlier, Marc Rich & Co., A.G., had supplied South Africa with more than 55 percent of its crude oil, as well as a significant amount of Jamaican alumina, in direct violation of UN trade sanctions. Manley and his party vowed to sever ties with Rich if they won.

Yet within a few weeks of his election, Manley's administration was under intense pressure from the International Monetary Fund (IMF) to raise more than $50 million in Western currency. When Marc Rich offered to come to the rescue with $50 million as a cash advance against future alumina production, Manley found it nearly impossible to turn him down. He told supporters at the People's National Party conference later that year, "The cash flow must be maintained." Without Rich's money, the currency would have been devalued, thousands of Jamaicans would have lost their jobs, and the economy and Manley's government would have been destroyed.

Other tales about Rich's trade deals surfaced as well. Just a year before the Ravenswood lockout, the Mexican government charged him with conspiring to sell copper from the state-owned Cananea mine at five percent of the world market price, in order to push the mine into bankruptcy so that he could then purchase it at "a bargain-basement price."

According to a former Citicorp oil expert, whether they were plundering state-owned businesses, violating the Iranian oil embargo and South African trade sanctions, or making profitable metals trades with Chile's Augusto Pinochet and Romania's Nicolae Ceausescu, Rich and Pincus Green "made a business out of doing business other people would not do." Their trading partners remained loyal. After Rich and Green were indicted for tax evasion in 1983, the Soviet newspaper *Izvestia* ran a front-page article defending Rich and accusing the U.S. government of a gross abuse of power.

At the same time that Marc Rich was America's most wanted white-collar criminal, he received more than $65 million in U.S. government grain export subsidies through his Swiss-based grain trading subsidiary, Richco. He used these subsidies to sell American wheat and barley at enormous profit in the Soviet Union, Saudi Arabia, and China. "No, it couldn't be," wrote *Washington Post* reporter Jerry Knight, somewhat tongue in cheek. "Not Marc Rich. Not the fugitive commodity trader. Not the one on the wanted poster. Not the one the Internal Revenue Service will give half a million to get its hands on. Not the one who renounced his U.S. citizenship and skipped the country to avoid taxes, wire fraud, and racketeering charges. Marc Rich, getting government subsidies?"

However, in a nation jaded by Watergate, the Iran-contra scandal, and the savings and loan crisis, these revelations made barely a ripple. A House agricultural subcommittee launched an investigation, and in October 1989 Richco was barred from participating directly in the USDA subsidy program. But, except for some bad press, Rich himself remained unscathed, having made hundreds of millions of dollars in the process.

Richco's U.S. offices were housed in the same offices as Clarendon, in Stamford, Connecticut. Jerry Knight discovered, "When you call information in Stamford for Richco Grain, the operator gives you the same phone number as when you call for Clarendon. And then Clarendon's switchboard operator asks, 'Richco? Do you want the grain department or the fertilizer department?'"

Hougan's interviews with former Rich associates and employees in the United States, Amsterdam, Paris, Zurich, and Geneva revealed Rich's extensive security apparatus. Many of those Hougan interviewed warned him about Rich's power and his willingness to use that power ruthlessly. One former employee in Switzerland warned him in particular never to go after Rich in Jamaica, because Marc Rich controlled the economy and Jamaica would be "the easiest, cheapest place to have you killed," a "piece of cake." Hougan learned that Rich had a retinue of private investigators who regularly spied on the lives and dealings of his friends and enemies alike. Nervous about being followed, he traveled with a rotating team of former Israeli soldiers, rumored to be armed with Uzis. As Mike Boggs explains, they learned that Marc Rich was "a rich thug, but a thug nevertheless."

Outside the circle of metals traders and wealthy celebrities in which he regularly traveled, Rich remained an enigmatic recluse. In his years in Switzerland he granted only one interview with the U.S. media, a 1986 meeting with reporter Shawn Tully from *Fortune* magazine. "In person Rich is soft-spoken and extremely shy," reports Tully. "Short and slim, his narrow face framed with slicked-back hair and long, bushy sideburns, he exudes an air of dour gentility. At times his reserve drops to reveal a lively sense of humor. Once, asked where he'd been traveling lately, Rich quipped: 'Not to the U.S. That would be a one-way trip.'"

Rich's way of life suffered little from his exile in Switzerland. He lived in a hilltop mansion with a spectacular view of the Zugersee (Lake Zug), filled with designer furniture and an extensive collection of paintings by Picasso, Miro, Braque, and Leger. He owned a luxurious chalet in the ski resort of St. Moritz, Switzerland, and a $9.5 million seaside estate in Marbella, Spain, that could house more than forty guests. He and his wife Denise held frequent lavish parties in Switzerland and Spain, inviting princes, oil sheiks, Fortune 500 CEOs, and European celebrities such as Placido Domingo.

In 1984, at his fiftieth birthday party at the National Hotel in Luzern, Rich's partners presented him with a ten-foot chocolate sailboat, along with a title to a real motorboat. His wife Denise, a rock-and-roll singer-songwriter whose "Frankie," sung by Sister Sledge, would become the number one hit in Britain within the year, performed two of her own compositions, "Don't Look Back" and "The Years Go By So Quickly." The highlight of the evening was a mock boxing match between a New York City policeman and a clown wearing the Marc Rich logo, refereed by another clown in judge's robes.

Despite these luxuries, Rich still lost a great deal from his exile. On his Spanish passport, he could travel freely from Switzerland to Spain. Yet everywhere else he risked being nabbed by Interpol police, bounty hunters, customs agents, or U.S. federal marshals and returned to trial, fines, and prison in the United States. There were several close calls. In 1986, he barely eluded capture at Heathrow Airport. "Rich apparently flew into England aboard Swissair," explains Craig Copetas. "Upon heading for the departure gate for the return flight to Zurich, he must have noticed that security teams were conducting complete checks of

identification and hand luggage. In a panic he scrambled to a public telephone and abandoned three checks payable to him (totaling 1.6 million
pounds) in the pages of a phone book. Then he boarded the plane and
flew home."

Most of all, Rich wanted to return to the United States. Though born
in Belgium and traveling under a Spanish passport, for Rich, America
was home. When his father died in New York in September 1985, he
could not attend the funeral, knowing that federal agents would arrest
him as soon as he entered the country. His elderly mother still lived in
the United States, and she was too ill to travel. Rich's wife Denise had
also returned to New York with their youngest daughter and filed for divorce.

"I want very badly to be able to go back," Rich declared in the 1986
Fortune interview. "I think about the U.S. every day. My mother is there
and my in-laws. It's a generous country that accepted my parents and
me. . . . I've made mistakes. I guess my reputation will never fully re-
cover. I've been portrayed in a horrible way, as a workaholic, a loner, a
money machine. It's not a true picture. I'm a modest, quiet person who
has never done anything illegal. What happened to me was an unfortu-
nate chain of events that hasn't shaken my faith in the U.S."

Rich had hired Washington lawyer Leonard Garment to try to work
out an arrangement with the U.S. government so that he could visit his
mother and daughter without facing hundreds of millions of dollars in
fines and a multi-year prison sentence. Garment had been Richard Nixon's
law partner in the 1960s and then followed him to Washington as White
House counsel. After Nixon's resignation, Garment had continued to
represent high-level Republicans, including Reagan's former National
Security Advisor, Robert McFarland, and Attorney General Edward
Meese. But according to Uehlein and Yeselson, Garment was "a charmer
who seemed more like a Democrat—a New Yorker, part-time musician,
Jewish—than a 'real' Republican." Despite his influential Republican
clientele, Garment had developed good relationships with Washington's
Democratic Party establishment, including organized labor.

Hougan had gotten to know Garment while interviewing him for his
book on Watergate. In early April, he set up a meeting with Garment,
George Becker, Paul Whitehead, Mike Boggs, and Joe Uehlein. The goal
was to leak the information they had collected on Rich and to signal
Rich, through Garment, about the worldwide pressure campaign the
Steelworkers were going to launch if he would not use his influence to
end the lockout. As they sat down in the meeting in Garment's office,
Becker was struck by an elaborate chart on the opposite wall.

I saw a big chalkboard that had all kinds of markings on it and arrows. A lot of them were stock symbols and arrows and like 300 mm with dollar signs on it. Millions, I guess. I don't know. I mean it was just jammed up with interlocking lines running everywhere. And I saw one thing on that thing that said CLAR written in there real small with an arrow pointing to it, and I just took a stab. . . . I said, "Leonard, how do you keep track of all the stuff that he owns?" And he said, "It's a son of a bitch, ain't it?" And he says, "And that's only a part of it." And he flipped it over and there it is on the back, Marc Rich. That's what it was, all Marc Rich. It was Clarendon.

For Uehlein, who with Yeselson had spent months trying to put together their own chart of the Marc Rich empire, it was a humbling experience. There across the room, too far for them to see in any detail, was a chart outlining everything they had been trying to prove for months, just out of reach.

During the meeting, Becker outlined the situation in Ravenswood, emphasizing their knowledge of Rich's operations and his interest in returning to the United States. According to Becker, Garment quickly replied, "The only thing I'm interested in is getting this misunderstanding cleared up and getting him back to the United States. He's not a criminal. He's done nothing wrong. He just happened to be extremely wealthy and the media came down on him." But when Becker told Garment that settling with the Steelworkers might help Rich return, Garment responded, "Let's be frank. There's not a goddamn thing that you can do, you or anybody else can do to help get Marc Rich back to this country . . . but I'm afraid there's a lot you can do to stop him."

The meeting ended with Garment promising to pass their message on to Rich. But as they left, he passed on an ominous message of his own: "Do what you have to do, but if you do, there will be blood on both sides of the tracks."

The time had come to go after Rich. Despite his wealth and power, Rich was vulnerable to a union pressure campaign because he wanted to get back to the United States and, as Garment said, the union could do a lot to keep that from happening. But, before launching a full-scale campaign, the union would give him a final chance to work something out privately. On May 31 Michael Boggs sent Rich a fax stating, "You might want to discuss the Ravenswood Aluminum case with us because it has implications that are much wider than the case itself and we'd rather settle it than have it develop to those points."

Rich sent back a fax of his own: "Dear Mr. Boggs, Thank you for your letter of May 31, 1991 requesting a meeting with me. Contrary to information provided by the USWA, I must advise you that neither I personally nor the Marc Rich group have any ownership or position in Ravenswood Aluminum Corporation." The fax concluded, "Against this background it would be pointless for me to meet regarding a topic over which I have no control nor involvement. . . . Sincerely yours, Marc Rich."

The time had come for the union to "let the dogs loose" on Marc Rich, as Michael Boggs called it. They needed to move quickly. With Boyle poised to buy out Bradley and Strothotte, the union stood in jeopardy of losing its leverage against Rich. They would need to take the fight directly to Europe, both to start the pressure on Rich and to cut off Boyle's funding source for the buyout, the same NMB Postbank that financed RAC's $50 million revolving credit line.

14

IN MARC RICH'S BACKYARD

By the end of May, the strategy team decided that within the month a union delegation would go to Switzerland and the Netherlands. Early in June, Uehlein and Hougan went to Europe to make arrangements with the Swiss metalworkers union (SMUV), the Swiss trade union federation (USS), and the Dutch trade union federation (FNV). Michael Boggs from the International Chemical and Energy Workers (ICEF) helped them set up meetings with the Swiss trade unionists and provided an interpreter. The International Metalworkers Federation (IMF) also assisted. From the beginning, the American and Swiss unionists collaborated. Explains Uehlein,

> They [SMUV] were consulted from day one . . . it wasn't "Here's our plan, we want you to do this, and this, and this." It was "Here's our problem, got any ideas?" And they did. And in fact we were even warned that the Swiss metalworkers wouldn't do much, that they don't understand strikes, that they kind of have a cozy relationship with the companies and the watchmakers and all that. But we found that to not be true either, they went way beyond. . . . I mean they assigned some staff full time. They arranged the press conferences. They paid for the luncheons. . . . And all this in the face of the company, the Marc Rich group, lobbying them not to do anything for us.

The Swiss union leaders took Hougan and Uehlein to Zug and introduced them to Josef Lang, a leader in the Green-Socialist Alternative Party and a member of the Zug city council. An anthropology professor at the University of Zurich, fluent in four languages, Lang was active in the teachers union at the university as well as local politics. He first entered Zug politics in 1982, when the Socialist Workers/Trotskyist group he belonged to merged with independent, feminist, and environmental parties to form the Green-Socialist Alternative Party. Five of the forty Zug council members were party members. Closely allied with the local labor movement and other social activists, Lang had spent his years on the city council actively monitoring and challenging the dominant role played by world metals traders in Zug. At the same time that Zug was a tax evasion paradise with a local government closely tied to financial and corporate interests, it also was a community where the left and Green forces had a strong political presence, garnering 30 to 40 percent of the vote. The Zug Kantonal chief of police came from the Green-Socialist Alternative Party and was very supportive of the Ravenswood workers' struggle. Thus, to Marc Rich's chagrin, he could do nothing to prevent mass rallies in his own backyard.

From Lang, Uehlein and Hougan learned a great deal about Rich and his life in Switzerland. To their delight, Lang "had been a Marc Rich watcher for years" and took them to his extensive basement library filled with information on Rich and other secret Swiss corporations that the Green-Socialist Alternative Party had been monitoring for over a decade. Lang had developed excellent contacts with reporters from American business publications such as *Fortune* and the *Wall Street Journal,* and was one of the primary sources for many of the articles about Rich that Uehlein and Yeselson had been poring over that spring.

He also confirmed what they had long suspected: Rich with his billions was impervious to financial pressure. Any profits he made or lost through his connection with Ravenswood Aluminum represented an insignificant piece of his global empire. Nevertheless, Rich was not without pressure points. Obsessed with secrecy and privacy, he would do everything he could to avoid attention from the media. Publicity was a threat to his ability to negotiate secret trade deals that danced on the edge of legality and corporate ethics. And most of all, publicity threatened his efforts to work out a deal with the U.S. government so that he could return to the United States without going to jail.

"He was too big to win against only because of losses he could have, because of the problems in Ravenswood . . . financial problems," explains Lang in his somewhat halting English. "But I think that his Achilles' heel was his image in the United States. I knew that the best manner to [pressure him] was put him another time in all the attention of the mass media. From the first moment I knew that it was the only tactic to win against Marc Rich."

Once Uehlein and Hougan described their campaign to Lang, he eagerly committed to doing everything he could to help. For the rest of the campaign Lang would be their key contact in Switzerland, teaching Hougan how to research a Swiss corporation, facilitating press conferences, and introducing them to unionists, legislators, and other activists to support the campaign.

Uehlein and Hougan found Zug a city of contrasts. Together in one city were Joe Lang and his socialist allies and the world's wealthiest metals traders. The quaint wood and brick buildings with red geraniums in every window box that once dominated the city were now dwarfed by the shiny new office buildings where Rich, Phibro (formerly Philipp Brothers), and 3M were headquartered. Rich's building, a six-story cube of blue reflecting glass, stood out from the rest. As described by a Swiss reporter, only at night would a passerby have any sense of what went on within.

The Rich headquarters in Zug, right next to the train station, is a creature of the night. The glass walls of the cube reflect the surroundings, making the building invisible during the day, when it takes on the appearance of the neighboring buildings. Not until the other businesses in Zug close, the attorneys and trustees put their briefs back in the safe, and the lights go on at Marc Rich, does the building become transparent. Until late in the evening, one can see men behind the glass walls sitting at computers and telephones, indicating in their way to passersby that their business extends beyond the borders of the small canton of Zug.

Partly out of curiosity, and partly to start the pressure, Hougan and Uehlein went to Rich's headquarters and asked to see Rich or Strothotte. Inside, the lobby was lavishly decorated, with a burgundy marble floor and thick salmon-colored carpeting. Artwork lined the walls, and soft jazz and popular music were piped through the public spaces. At the center sat a large semicircular table with four or five receptionists and secretaries. Uehlein and Hougan spent forty minutes trying to convince a secretary to let them in to see Rich. Uehlein describes the scene:

To get beyond that . . . it's like going into the FBI. You have to have a special card that goes into one of these stands that comes out of the ground and you go through a door. We never got in past that. But we're dealing with this one secretary . . . She would go and talk to him and come back and say "He's not going to see you." And then I would lobby her harder and she'd go talk to him again, and come back, "He's not going to see you.'" Finally I had one of these info packets that Gary Hubbard had put together for us, had the chart, had a lot of the stuff, it really was kind of provocative. And I gave it to her with my card and I said, "Show him this chart, show him what we're doing, and then see if he'll see me." So she goes in, and she's in there for about a half an hour this time. When she finally did come out, she says "Look, Mr. Rich is very busy. If you'll tell us where you're staying tonight, he'll contact you later." And I laughed. I said, "Would you do it if you were me?" And she just kind of looks at me.

Previously warned that it was not safe for them to stay in Zug, Uehlein and Hougan were not willing to trust Rich's intentions in asking them where they were staying. So they left without seeing him. But they did not leave empty-handed. While the secretary was talking to Rich, another secretary had answered the phone.

Another secretary picks it up and she's repeating everything this guy's telling her, name, address, phone number. And it's a reporter from *American Lawyer* magazine, that's the name of the magazine in New York . . . all I did was get his name and phone number. And I went out and called him once we were snubbed by Rich. . . . And he was trying to find out why Robert Thomajon, a named partner in a prestigious New York firm, would leave the firm and go to Zug, Switzerland. He wants to interview Thomajon. And we didn't know Thomajon was in Zug, Switzerland. We didn't know Robert Thomajon had become a top aide to Marc Rich. We knew from the loan agreement that the firm, Milgram, Thomajon, and Lee, did that deal. And that's when it clicked. And we thought, aha, Robert Thomajon is here in Zug working directly for Marc Rich. His firm handled the Ravenswood buyout.

The next day they fit another piece into the puzzle when Hougan traveled to the small town of Hergiswil to search for the headquarters of Ridgeway Commercial, the company that had financed RAC's original leveraged buyout of the Ravenswood smelter. Yeselson had discovered the Hergiswil address for Ridgeway in the loan agreement document in the Ripley courthouse. Nestled high in the Alps, Hergiswil is a town of picturesque chalets and restaurants circling a small mountain lake, an unlikely site for a corporate headquarters. One of the chalets, the Alpena Treuhand, was listed as the Ridgeway address. To Hougan's amazement, a brass plate by the door listed another thirty corporations with the same address. Yet inside, for all these thirty corporations he found just two or three staff, dressed casually in jeans, T-shirts, and sweaters. As Joe Uehlein later described it, "Not only was Hergiswil the quintessential Swiss Alps town, it was the quintessential secret Swiss corporate front, this little chalet."

The office staff were clearly surprised to see Hougan and even more surprised when he started to ask questions about who owned Ridgeway

Commercial. To Hougan's delight, completely unprompted, one clerk blurted out, "Marc Rich will have to get back to you." Rich never did get back to him, but Hougan had gotten the information he wanted. Without even mentioning Rich's name, he had been told at corporate headquarters that Marc Rich was the person to talk to about Ridgeway.

After Switzerland, Uehlein and Hougan moved on to the Netherlands to work with the Dutch trade union federation and Hans Noten of the bank workers union, to set up the meeting with NMB Postbank officials. Noten, who was the union liaison to the NMB Postbank board, was willing to help the Steelworkers set up a meeting with the bank, but was hesitant to do anything that would jeopardize the good relationship between his union and the NMB. He also cautioned the Americans that a street demonstration outside the bank would backfire, warning that "such things weren't done by the unions in Holland."

Uehlein and Hougan returned to the United States on June 7. Uehlein pushed from the beginning to send a delegation of locked-out workers to Europe, rather than just a few international staff and officers. He knew that the most effective way to tell the story of the lockout and what it was doing to workers' lives, their families, and their community was for rank-and-file workers to take their story to Marc Rich's doorstep and tell it themselves, in their own voices. Despite the cost of sending a negotiating team of five people, and despite the possibility of flak from other international officers, Becker agreed.

The local delegation was due to leave for Switzerland on June 22, yet at this point, except for Becker, Uehlein, Hougan, and Yeselson, few knew of the plan. In less than three weeks arrangements needed to be finalized in Switzerland and the Netherlands; press materials had to be prepared in English, German, French, and Dutch; passports and airline tickets had to be obtained for the entire committee; and the union delegation needed to be prepped for the trip. Except for some stints in the army, most had never traveled much outside the Ohio Valley, and certainly not to Europe. Some, like Mike Bailes, had never even flown before.

When the news broke, many in the union, particularly Chapman and Ellswick, seriously questioned the value of chasing Rich across Europe, given both the expense of the trip and the need to focus staff and financial resources on the end-user and NLRB campaigns back home. Eyebrows were raised at the Steelworkers building in Pittsburgh, and many rank-and-file members were skeptical. To them Boyle was the villain. But Becker and the strategy team knew that pressuring Rich was crucial to winning at Ravenswood and they were prepared to defend their decisions.

One week before they left, the trip almost fell apart. Shortly after Hougan's and Uehlein's visit in early June, the presidents of the Swiss trade union federation and the metalworkers union sent a letter to Rich, once again asking him to intervene in the dispute. The letter was also signed by five members of the Swiss parliament. As Uehlein tells the story, they got an immediate letter back from Rich warning, "[D]on't believe those American unions. I'm not connected with that thing. They're pulling the wool over your eyes." Uehlein continues, "So the Swiss Metalworkers sent us an urgent fax a week before we're about to leave. We're all ready to go. They said, 'Everything's off unless you can prove it.' They'd never asked us for proof. They took us at our word. Now they're saying, 'Unless you can prove to us what you've been saying, we can't put our neck out.'"

Uehlein, Yeselson, and Hougan rushed to assemble documentation on Rich's connection to Ravenswood. Much to their relief, within days of express-mailing the materials, they received a positive reply. "You've proved it. We accept it," it said. "Thank you for the documentation." And from that moment on, despite continuous pressure from Rich, the Swiss labor movement never wavered in support of Local 5668.

On June 22, the local delegation met Joe Chapman at the Charleston airport and flew to Dulles, where they were joined by Hougan and Uehlein for the long flight to Paris and then Zurich. The trip itself made many of the workers uneasy. The food was strange. The language was strange. And several of them, like Dan Stidham, were white-knuckle flyers, taking the trip only because they were willing to do whatever they could to get their members back into the plant.

In Zurich Joe Lang met them at their hotel and shepherded them around during their stay in Switzerland. Lang could tell how uneasy the Americans were to be in a foreign environment. "The reaction of the rank-and-file workers in Zurich, or in Zug, reminded me of the same sort of habits of my father or my brothers when they come visit me. Like being in another world," explained Joe Lang. "Not people who are used to living or moving outside their life worlds. Not people who are used to speaking with people who don't know English . . . or speak it like me."

But Lang was struck by the Ravenswood workers' commitment. He found them very different from their counterparts in the Swiss metalworkers, most of whom had a detached, professional relationship with their union. For Stidham, McDowell, and the others, the union was their fatherland, their "Lebenswelt," their world of life. And it was this commitment, and the sincerity that went with it, that Lang felt made them such effective emissaries for their cause. For their part, the Local 5668 members took to Joe Lang right away. They were amazed at his wealth of knowledge about Marc Rich and moved by his dedication to them and their campaign.

Their first action was a press conference outside Rich headquarters in Zug. The Swiss metalworkers had publicized the event, which was well attended. Throughout the crowd signs could be spotted declaring the solidarity of Swiss trade unionists with the "American Aluminum workers." Uehlein and Chapman described the lockout, the NLRB complaint, the OSHA charges, and the connection between Rich and Ravenswood. Uehlein declared, "If we did not believe in the relationship between Ravenswood and Marc Rich, we would not have come to Zug. . . . It is now up to Marc Rich to put the cards on the table and prove to the public that he has no stake in Ravenswood."

Then Dan Stidham, after introducing himself as "a union man all my life," told the crowd what Emmett Boyle, Marc Rich, and the lockout meant to the workers and their community. He ended his remarks:

> But we won't go away, and neither will all the people around Ravenswood who are supporting us. They don't like scabs any more than we do. And they don't like absentee owners lining their own pockets at the expense of the standard of living of entire communities. And that's what's happening. Without paychecks, we've had to struggle to support our families. We're not buying all the things we need, and small businesses, ones that we've patronized for years, are really hurting. They're having to cut back, too, and lay off workers. But RAC doesn't care. Not as long as it can sell aluminum and hire scabs. That's why we're here, to get some help from our union brothers and sisters in Europe, so we can help our union brothers and sisters back home.

Representatives from the Swiss unions and a Swiss legislator also spoke. The union delegation, followed by the media, then marched into Marc Rich's headquarters to request a meeting. Inside they confronted the same phalanx of secretaries that Uehlein and Hougan had run up against a few weeks before. They brought with them a petition from the union federation, signed by several members of the Swiss parliament, including the president of the Swiss Social Democratic Party.

To no one's surprise, Marc Rich refused to meet with the delegation. But before they left the building, Charlie McDowell had himself photographed pointing to the Clarendon sign, just below the sign clearly spelling out Marc Rich & Co., A.G. Outside the building, the ever garrulous McDowell then held court with the Swiss reporters. "As broad as

a giant and with a handshake difficult to forget in a while, Charles Mc-
Dowell, a perfect example of a steelworker, is 57 years old, has been married for 37 years, and is a proud father of three grown children," extolled
a Swiss reporter. "However, since October 31 for him . . . the world broke
apart. . . . He has traveled together with a union delegation to Zug in order to speak with Rich. 'I would have told him he should sit together
with us at the table and talk.'" The event got six minutes of coverage on
Swiss television that night, including a segment that the stations had
filmed at Fort Unity the week before.

The small delegation of Steelworkers had accomplished in one day
what Joe Lang and others had been trying to do for years. They had the
citizens of Zug and Switzerland questioning whether Marc Rich, the
great philanthropist, was good for their country and their community.
"It was the first time that a campaign against Marc Rich had moral support from the majority," Joe Lang explained. "All our campaigns [fo-
cused on Rich's breaking the] embargo against South Africa, the deals
with Pinochet and Ceausescu, the deals with Pinochet and China, their
deals with Iraq and Iran, their profiting from embargoes. All [those] campaigns didn't have this impact, of Ravenswood . . . because it was easier
to understand what was going on . . . what's visible, a thousand seven
hundred people locked out."

While these actions were going on in Switzerland, the Steelworkers
were holding a legislative conference in Washington, D.C. At one point,
Joe Uehlein and Charlie McDowell were hooked up by telephone to give
a live report on their activities in Zug. The USWA delegates then
marched the ten blocks from the conference to the Swiss embassy. At the
head of the line, Lynn Williams and George Becker wore signs printed
"U.S. Criminal Marc Rich Safe in Switzerland" and "Swiss Should Investigate Marc Rich."

While Jim Bowen led the chanting delegates outside, Becker and
Williams were admitted to the embassy to deliver a letter to the ambassador. The letter called on the Swiss to reevaluate Marc Rich's continued
residential status in Switzerland and to institute a parliamentary inquiry
into the Marc Rich group, including allegations that the company had
violated UN trading sanctions and Swiss law and had had a "corrupting
influence on the Swiss political process."

In Europe the union delegation moved on to Amsterdam to meet
with NMB Postbank officials. Their hope was that, by presenting their
story to NMB, they could disrupt Boyle's deal to buy out the other
owners of the plant, thereby demonstrating their power and effectiveness to both Boyle and Rich. The meeting was arranged by Dutch

unionists, including the union representing NMB employees. As Joe Uehlein describes it, the meeting was a great success:

> They went in there and this banker, he was a member of the board, part of the management team at the bank, was so struck by these guys for a lot of reasons. . . . [They] exuded West Virginia. I mean, they were genuine. And that just came out of every pore in their bodies and it struck this guy in a way that a Hougan or me or anybody else could not have. That would have been more like a business relationship, more of a formalized kind of thing. And he wanted to know every one of their stories. And he sat down at the table, and after we went around and did the introductions, he then asked Dan Stidham, "How are you feeding your family? How are you paying your medical bills?" And goes to the next one, "How are you, what is this lockout, I don't understand this, how does this occur?" He couldn't believe it. And they felt really good . . . that they got an hour with the guy and that he was genuinely moved. . . . They left feeling very confident that he was going to do something.

The bankers were particularly interested in the NLRB complaint and the millions of dollars of back pay liability accumulating each month. Although the bank officials made no commitments that day, the union members left the meeting convinced that they were reconsidering their loan to RAC. Later that day the union held a well-attended press conference.

After the press conference, Jim Hougan took a side trip to the office of corporate registrations to research the connection between Rinoman Investments and Marc Rich. One more piece fell into place. Not only did the corporate papers list Willy Strothotte as Rinoman's chief officer, they also revealed that NMB was the registered agent for Rinoman, despite the bank officials' denials earlier in the day that they knew of any connection between Marc Rich and the Ravenswood plant. The next day Uehlein received a phone call from Hans Noten reporting that the bankers were very impressed with the workers' presentation and was prepared "to take an initiative with RAC."

On June 29, as the delegation headed home to Ravenswood, the members were weary but excited. "I think we all thought that we were on to something," Dewey Taylor recalls. "That if we could direct enough attention to Marc Rich . . . he's the one guy that with a word or a snap of a finger or something could end it all. We thought he was that powerful in the scheme of things. That he could just put his finger on the button, you know."

They were greeted by a full-page advertisement in the *Jackson Star News*, paid for by RAC. Headlined in large letters "How Certain USWA Officials are $pending the $ummer," the ad read, "We thought you might like to know how some United Steelworkers of America (USWA) members are spending the summer and a significant amount of money. $Visit Switzerland . . . $Tour the Netherlands . . . $Tour the Nation's Capitol . . . $Overnight Trips . . ." Chastising the leadership for spending the members' money and spreading "misinformation and lies" about RAC with "no effect on the labor dispute," the ad concluded:

> It takes money to travel, especially to Europe, and those paying union dues may want to think about who is picking up the tab. Perhaps the slogan is wrong. Maybe it should be "Join the USWA and, if you are the right USWA member, see the world." If you are the wrong USWA member, you are presently out of work because the right members never gave you an opportunity to vote on the final proposal Ravenswood Aluminum Corporation submitted to your negotiating committee.

Fortunately for the local, the company had little credibility with the rank and file. A few grumbled about the "European vacations," but, as one local member reported in the *Lockout Bulletin*, "they might fool someone that knows nothing about this situation, but fooled nobody that knows the track [record] of truth coming from RAC."

At the weekly informational meeting and picnic on the Tuesday after they returned from their trip, the negotiating committee regaled the large crowd with stories of the trip to Zug, the meeting with the NMB Postbank, and their newfound friends and allies in both countries. Suddenly Marc Rich seemed much more real and much more vulnerable to their campaign. Lab worker Betty Totten would later tell a reporter from *The Nation*, "I spent my entire working adult life in the plant and I was just thrown out. Up to then, my life was set. . . . I never dreamed I'd be [sitting here] worrying about someone like that. And I bet he never dreamed he would have to worry about people like me."

15

THE TIDE TURNS

As the summer of 1991 heated up, the European campaign had little impact in West Virginia. The Tuesday night suppers continued, the Assistance Center paid bills and distributed food, and members of Local 5668 walked the picket line. The union was holding in Ravenswood.

On July 10, the locked-out workers got to preview a video, "The Battle of Fort RAC," produced by the USWA to use in the campaign. Featuring interviews with Dan Stidham, Bill Doyle, Charlie McDowell, and Jim Bowen, the video reviewed the problems with safety and health, the deaths in the plant, and RAC's plan to force the union to an impasse. George Becker outlined the "coordinated campaign strategy" the union was using. Already shown on Swiss television during the union delegation's visit, the video was a powerful statement of the union's commitment to victory. It showed Bill Hendricks pointing to a picket sign and declaring, "Just like that sign says, 'One day more.'"

In its ongoing investigations, the union discovered that Rich's Clarendon Ltd. had recently won a bid to supply $1.3 million worth of nickel and $4 million of copper to the U.S. Mint. This was above and beyond the $23 million of nickel, zinc, and copper that Clarendon had sold the Mint in the two previous years. On July 11 a delegation of eight locked-out workers joined local Steelworkers from USWA District 7 for a noontime rally in front of the Mint in Philadelphia. The small rally was poignant. Here was a fugitive from justice receiving a lucrative government contract—at the same time that he was taking away the livelihoods of thousands of workers in a small West Virginia town. The news flew across the business press both at home and abroad.

But the international intrigue seemed not to matter when the NLRB announced on July 19 that an unfair labor practice complaint would be issued against RAC within the week. The complaint alleged that RAC failed to bargain in good faith. Contrary to the company's claim, no valid impasse had existed on October 31, 1990; therefore both the lockout and RAC's unilateral implementation of its final contract offer on November 29, 1990, were unlawful.

A trial in front of an NLRB administrative law judge was set for September 23, 1991, where both sides would have an opportunity to present their cases. Yet, as the *Parkersburg News* headline suggested the day the complaint was issued, "History Seems to Favor Steel Union." Reviewing NLRB decisions over the past several years, the paper reported that four out of five complaints brought by the Board were upheld either by administrative law judges or by the full Board itself.

In an emotional gathering at Fort Unity, Jim Bowen told an ebullient crowd, "It is the NLRB that will now make the issues with the company, along with our support." Interrupted by applause, spontaneous cheers, and firecrackers that almost prevented him from speaking, Bowen went on, "It is now the government that says RAC violated the law, not the union."

True to form, Joe Chapman was more restrained. "I'm very proud of every one of you," Chapman told the crowd. "But I'm always cautious, and I'm going to be cautious in this. This is one giant step toward what we [USWA] committed to you was going to happen. You're going to get your jobs back and the scabs are going down the road and we're going to get this community back to normal. But don't get impatient with us." He told the assembled members and their families, "Remember, it may not happen overnight. Keep cool, keep the faith, and stay off the roads."

But the union was celebrating more than a simple legal victory. NLRB Regional Director Ed Verst was clear that, if the complaint against RAC was sustained, the company would be liable for back pay and benefits for all seventeen hundred workers back to October 30, 1990. Verst said that the Board was also considering the union's request for an injunction that would put workers back to work immediately.

In Pittsburgh, Rich Brean and the rest of the legal staff were ecstatic. To think that this was the case they had literally saved from the mail room. "I remember we all went over to Palmer's Restaurant, right where their [RAC's] law firm is, Polito and Smock," Brean remembers. "And I remember I ate a ham steak and home fried potatoes, just this victory meal, that we had gotten the complaint."

The union hoped the Board complaint would bring RAC back to the bargaining table, to avoid the tremendous liability that the judge's decision might bring. Talks were scheduled for July 30, under the supervision of federal mediator Carmen Newell. But if the Steelworkers thought that the NLRB would soften RAC's resolve, they were mistaken. Talks fizzled after less than two hours. RAC still refused to bargain. Joe Chapman reported about the session, "They restated their commitment to the replacement workers and basically refused to discuss any of the outstanding issues." Rich Brean added, "We really thought that if . . . they're telling you a hundred million dollars your corporation's in jeopardy, that somehow this would bring reason. But not to these people."

The bad news for the Ravenswood Aluminum Company didn't quit. On August 2, West Virginia Senator Robert Byrd announced that OSHA would be ordering a "wall-to-wall" inspection of the Ravenswood plant. While inspections of specific complaints, such as the one RAC went through in June, were routine for OSHA, wall-to-wall inspections were rare, especially given that Republican administrations had been underfunding OSHA for the previous decade.

The Steelworkers had been hammering away at health and safety issues for months in the press, to the public, and to government officials. Jim Valenti explains how their approach evolved.

> My thinking went from more of a regulatory, traditional thinking of the safety and health department—go in, get an OSHA citation, file party status, make sure the citations stuck—to more of a corporate campaign thinking, and doing things that generate public support and interest. That's when we started playing with the idea that this aluminum was made with blood. . . . That's where that whole public campaign came from, and we started contacting the council-people and the politicians and it was pretty hard for them to ignore it because the citation was so large, and there were five deaths in sixteen months. It was compelling.

RAC's refusal to admit OSHA inspectors in May had provided the Steelworkers with a perfect opportunity to appeal to their political leaders, including Senator Byrd, who played a major role in obtaining OSHA's wall-to-wall inspection.

The company did their best to prepare the workforce for the inspection, while steadfastly clinging to their alleged commitment to health and safety. Management told the salaried and replacement workers: "It is common in this type of inspection for OSHA to find some violations and, with a facility as large as ours, we expect this to be true at RAC. The inspectors may, in fact, discover a number of minor problems. However, we are confident that our facility has no serious health or safety hazards and that the OSHA visit will be positive and productive. Where we work is relatively clean, safe, and hazard-free."

The inspection began on August 5, 1991. Because the Steelworkers, despite the lockout, were still the official bargaining agent for RAC workers, Bill Doyle, as the union health and safety representative, was allowed to participate in the walk-through phase of the inspection. It was RAC's worst nightmare. Bill Doyle, the union vice-president, the man who had gotten this OSHA action started, was going to be walking around inside their plant. Doyle knew more about RAC's health and safety problems than anyone. More than that, the union would have its first opportunity to observe what the plant looked like when run by a scab work force. The scabs would also have their first opportunity to see the union in operation, in their workplace.

RAC managers did everything they could to cover up potential safety problems and to restrict the OSHA inspector's and Doyle's movements. They also tried to make the experience as uncomfortable as possible. Doyle relates, "I was with one of the OSHA inspectors, and they [the management] wouldn't let us get a drink of water. I asked them three times. For three hours they wouldn't let me get a drink of water."

But the company's best efforts to hide information from Doyle only backfired. At one point the company switched crane operators in the middle of an operation. But Doyle spotted the ploy and "they got cited for false access to the crane. They got cited for about ten citations right there because they did that, just to keep me from seeing who was operating that crane. Dozens of things like that happened."

Doyle was frustrated by the company's attempts to clean up their practices in advance of the OSHA team and to shield the inspectors from the worst violations. He had hoped that the inspectors would be more aggressive and thorough and would take his recommendations and input more seriously. He was especially disappointed that the inspectors allowed RAC to steer the tour away from casting, "the highest injury department" in the plant.

Although the penalties resulting from the walk-through would not be announced for months, still the inspection generated a great deal of negative publicity for RAC. More than that, in combination with the NLRB charges, the OSHA inspection had been a very bitter pill for RAC management to swallow. It was a shot in the arm to the rank and file, reinforcing the sense that the tide was finally turning.

Meanwhile, Uehlein, Yeselson, and Hougan continued to follow all leads. They knew that Emmett Boyle had gotten divorced, and Yeselson knew that divorce proceedings often contain important financial information. Particularly given Boyle's connections to Clarendon and Rich, they continued to pursue this angle.

Yeselson had been charged with tracking down the Boyles' divorce papers. But after a search of every county courthouse in the area, he came back empty-handed. As he later would recall, his mistake was searching in "straight alphabetical order." Boyle's divorce records had been misfiled.

Hougan was in Wheeling on other business and decided to search once more at the county courthouse. This time he found the file under "Robert Emmett Boyle." Without much ado the clerk handed him a thick file. "I was thinking, well, why was this supposed to be so hard? I think it was about an inch thick, this sort of file we're talking about. And in there were a number of pleadings that were interesting for what they said about the relationship between Ormet and Ravenswood and Emmett Boyle's participation. . . . There was also, of course, an order in there saying that whatever the number, something like nine thousand pages or whatever, of materials were not to be made available." Hougan made copies of eight or ten pages, along with detailed notes which he passed on to George Becker. Becker got very excited and sent Hougan back for the entire document. This time a different clerk was on duty, and she refused to copy the file.

Given his experience as an investigator, Hougan was not about to let that be the final word. He was directed first to Judge Callie Tsapis, who was on vacation at the time. Hougan went to visit the judge who was filling in for Tsapis, and got him to order the clerk to copy the entire file.

The divorce proceedings papers Hougan was handed that day were a gold mine. They documented that, in 1986, at the same time Boyle had negotiated an "equality of sacrifice" provision with the local union at Ormet, he had received a secret payment of $500,000 from the former owner of Ormet. "Boyle sat across the table and told us he would reduce his salary, hoping to pave the way for union wage and benefit concessions. Then he turned around and cut this secret deal," explained Jim Bowen. "As far as I'm concerned, he betrayed the trust of our members at Hannibal, and we're not going to sit still for it. I've directed our lawyers to file a lawsuit charging that these hidden slush fund payments violated the letter and spirit of the commitments made to the union in the 1986 negotiations."

USWA attorney Paul Whitehead suggests that Boyle's betrayal went far beyond these payments:

If you really put it together, it suggested that Emmett Boyle, one of the things he got in return for his so-called "equality of sacrifice," in return for his so-called pay cut, was one-fourth of the stock in the Ormet Corporation, or its parent, to be specific. By 1989, his co-owner, Thomas McGinty, sold that stock for $39 million. So in return for a pay cut, he gets a chunk of stock that's soon worth $39 million. It just so happens that our concessions, during that period, were also $39 million. It was always my belief that he used the deception of our union at Ormet as the way of building up his personal worth and . . . his prestige in the business community to launch his involvement at Ravenswood.

The file also shed additional light on the relationship between Rich and Boyle. At the time of the takeover, a front company called the "Really Helpful Corporation" was the largest owner of Ormet. The divorce papers revealed that the offices of this company were located at the offices of Rich's law firm, Milgrim, Thomajon and Lee. Once again the union's research showed that Boyle's power and wealth were primarily on paper and subject to Marc Rich's control.

Boyle was furious about the union's release of information from his divorce. "My privacy has been violated as part of a harassment campaign by the USWA International," he fumed. He went to court and got a gag order preventing the union from holding a press conference to further publicize the information about him, and filed a contempt complaint against the USWA, Hougan, Whitehead, and Bowen for willfully violating the confidentiality of his divorce proceedings.

In late August the union was stung by the NLRB's decision not to seek an injunction ordering that RAC bring union members back to work. While Board officials were careful to say that this decision was unrelated to the merits of the complaint against RAC, and the union knew that these kinds of injunctions were quite rare, the decision dashed the union's hopes for a quick legal resolution. They would have to wait for the NLRB trial to begin.

In the interim, more than four hundred RAC workers and their families traveled by bus to Washington, D.C. as part of a national labor rally and protest, "Solidarity Day II," called by the AFL-CIO. They joined close to half a million workers from across the country rallying for universal health care and the passage of Senate Bill 55 to prohibit the use of permanent replacement workers—a topic that members of Local 5668 knew only too well.

August 31, 1991, was a sweltering day in Washington, already in the 90s by mid-morning, with an oppressive humidity settling down over the city. Water was short and a number of marchers were overcome by the heat. True to form, Dan Stidham helped out. "I dropped out of the parade to assist the ones who were affected by the heat stress and never made it back in line. I never even got to hear the speakers."

Dewey Taylor, who was increasingly taking on the role of spokesperson for the local, was one of fifty-four speakers selected to make brief remarks from the main stage at the foot of the Capitol, above the hundreds of thousands of marchers filling the Mall. Taylor didn't get onto stage until 4:00 A.M. but, in his folksy down-home manner, he told the story of the Steelworkers' struggle in West Virginia. Buoyed by the show of solidarity, Stidham remarked as they were leaving, "We left our mark on Washington and hope it will open some eyes around the country."

The next day another fatal accident occurred. As a tractor-trailer navigated a curve on Interstate 64 near Huntington, West Virginia, a five-ton coil of Ravenswood aluminum fell off the back, crushing a car that followed. The passenger was killed and the driver was seriously injured. RAC spokespersons quickly defended the company: "It is RAC's understanding that the shipment was taken to a warehouse and unloaded; then on September 3, the coils were reloaded by the trucking company for shipment to the Southwest." But given their troubles with OSHA and the NLRB and the public's waning support, this accident was the last thing RAC needed.

The company did its best, however, to stay on the offensive. That same week, RAC attorneys filed suit against the Steelworkers for violating the West Virginia Trades Secrets Act. The suit was based on the continued use and publication of information from the Price Waterhouse Strategic Plan that the union had obtained in December 1990. "It is the right of every corporation to develop a strategic plan and to expect that this confidential information will not be distributed to the public, which includes competitors," Debbie Boger asserted.

To the union, this charge seemed like old news. Charlie McDowell called the suit "a smoke screen." Rich Brean had signed an agreement with RAC in March that the union would not use information from the Strategic Plan, and Uehlein, Yeselson, and Hougan worked doggedly over the ensuing months to find secondary sources confirming the information in the plan. They had gathered a great deal of new information since March, including Boyle's divorce documents and Marc Rich's connection to the U.S. government through his grain subsidies and sales to the Mint.

RAC continued this legal press two days later, adding a hundred more charges to the RICO suit and naming six additional individuals. That same day the company filed thirty-five new charges of violence with the NLRB, alleging that "union members harassed and threatened RAC employees, threw rocks and other objects passing the Local 5668 Union Hall on Route 2, and placed jackrocks on the road near the union hall." Local leaders saw these allegations as fraudulent, but, said Charlie McDowell, "Nothing [RAC] does really surprises me. Anything to take your rights away from you. I guess they're trying to keep us in court."

Despite RAC's legal actions, the union continued to gain ground. On September 21, the USWA District 23 conference in Charleston feasted on a banquet prepared by the Women's Support Group and gave more than $21,000 in donations, along with a ton of food for the Assistance Center. The union got another boost when the Stroh Brewing Company announced that it would no longer use Ravenswood aluminum. Stroh's had been an early target of the thousands of "bloody-can" leaflets distributed across the country, and Becker had written to Stroh's executives. In a letter to Becker, a Stroh's vice-president wrote, "[We] believe that we have taken the appropriate action to satisfy your previously stated concerns over the use of Ravenswood aluminum." Becker recalls, "It was a tremendous victory at just the right time. It convinced us that our strategy was sound." The months of tracking trucks, writing letters to CEOs, and handbilling public events were beginning to pay off.

But the real excitement in the fall of 1991 was the trial of the NLRB case. Several procedural breaks helped the union. One was Carol Shore, the Board agent assigned to develop and argue the case against RAC. The union also lucked out in the selection of Bernie Ries as the judge. Ries was known as one of the best administrative law judges in the region. According to Rich Brean, not only was Ries "five times smarter than the average Board member," but he had the intellect and curiosity that allowed him to see cases in a broader perspective. In addition, Ries had been the judge in the very case, *Stores Communication,* that Brean and Shore were using as the precedent to argue that the lockout was unlawful. When the lawyers heard which judge had been assigned, recalls Brean, "we just started high-fiving it. . . . to have the judge from that case assigned to this case. . . ."

The trial had been moved from the usual hearing room to the Charleston City Council chambers because so many people wanted to attend. Brean describes the setting:

It's a simple but dignified room. It's wood-paneled. And Ries was up in front. Absolutely crammed full of interested spectators. Standing room only, with crowds in the hallway. . . . And I think it was very much a social event, an historic event. . . . Not only were people divided by the labor dispute, but they were divided by generation. There were the older, something like average age 56 workforce, very much in attendance. Almost as if they were in church—so well-mannered . . . and then from time to time, a younger cohort of scabs, there to observe. There, we thought, at the behest of the company . . . with the hope of fomenting something by the other side. And they brought them in from the plant and they stood against the wall, literally smirking, twenty- and twenty-one-year-olds.

Union lawyers Rich Brean and Stuart Israel, whom Brean had brought in because of his talent for cross-examination, had worked day and night with Carol Shore, preparing their case. "Brean was driven to succeed in this case," Israel remembers. "And he was indefatigable in his efforts. He drove the rest of us, too. And I'm a workaholic and he made me feel like a lazy bum. I remember some miserable Sunday meetings with him in Cincinnati." "Carol Shore has two little kids at home in Cincinnati. She was working eighty- and a hundred-hour weeks, week after week in West Virginia. Away from her family, relentless," recalls Brean.

There were few surprises the first day, as RAC and the NLRB made their opening statements amidst boxes of files and stacks of legal pads. RAC attorney David Laurent argued that the union had been "utterly inflexible" in contract negotiations and had bargained to impasse. He went on to suggest that, because workers did not show up for work at the 8:00 A.M. and 4:00 P.M. shifts on November 1, 1990, the union was not locked out but, in fact, on strike. Carol Shore countered that the union had shown great flexibility in bargaining, including dropping the metal price bonus, and that RAC was inflexible, refusing to discuss safety and health at any time during negotiations. She argued that, because workers had been sent home in the early hours of November 1, the company had indeed locked them out.

These were the exact positions the company and the union had taken for many months in the press, in front of the West Virginia unemployment commission, and with the citizens of Jackson County. But this clearly would not be a simple case. Judge Ries told the press after adjournment the first day, "I had thought the case might take a couple of weeks. From the opening statements, I imagine it will take a lot longer."

Slowly building the union's case against RAC, the following day Bill Hendricks and Larry Monk testified that they were turned away from the plant on November 1, 1990. That afternoon Joe Chapman took the stand. Chapman would be the primary witness for the union's NLRB case. Unions often rely heavily on the chief negotiator, but typically a number of local union officers and committee members are called to testify, to make the union case more credible. Between Brean, Israel, and Shore, this crack legal team could have called as many witnesses to the stand as they wanted. But the lawyers had decided to build their case around Chapman. They were taken by his quiet, cautious demeanor, which they felt would be highly effective in front of Judge Ries.

Their decision was not without political fallout. The local committee members were itching to testify, particularly Charlie McDowell. They had been fighting the good fight in the trenches against RAC and wanted their day in court. Yet Brean and the rest of the legal team knew from experience that local union officials often play to their members and to their own egos, which is not necessarily helpful in presenting their case in court. They were particularly concerned about McDowell, who kept pushing to be put on the stand. Recalls Brean, "McDowell came up to me and said something like, 'You sure lost this case, we lost this case. All Joe did [on the stand] was surrender and surrender and surrender. He wasn't even tough. All he was doing was backing up, making concessions.'"

Chapman was not comfortable being thrust into the limelight. He already had a tremendous weight on his shoulders as the day-to-day staff person responsible for the lockout. And, while he was a very experienced union staffer, he did not have much trial experience. For months before the trial, in addition to taking his deposition, Brean, Israel, and Shore worked with Chapman to get him ready to testify. Despite his discomfort with taking center stage, Chapman agreed with the lawyers' strategy. "Sometimes we've made mistakes and not thought things through. We've done a lot of things for popularity purposes. When you get in a death struggle, you better do what's right, and the hell with all the politics of having local union people up there. You answer for that later."

Chapman's testimony went flawlessly. According to Brean, "Her [Shore's] direct exam of Chapman was the best direct I've ever heard. It flowed smoothly. It was so well prepared. He was comfortable with her. Probably more comfortable with her than he was with us. Because Israel and I both are a little crazy. Israel is calmer than I am. Carol Shore is much calmer than either of us."

The cross-examination by Laurent was tougher, but Chapman bore up under the pressure. Stuart Israel recalls, "Joe was kind of the focal point of all this. He's fighting the company, he's the star witness, and he's fighting his own people, who are criticizing every step he makes. And he carried it very well. Held up and did his job and I thought he was an excellent witness."

More important than what the lawyers felt about Chapman's performance was how it played in court. Brean was confident that they had won the case on Chapman's testimony alone. "I say that because you can read body language from Ries. . . . Because you can read a judge. Because when your witness testifies you don't look at your witness, you look at the judge. And Ries is going to credit the guy."

The company opened their case, calling Al Toothman, manager of labor relations at RAC. Toothman insisted that the preparations for bargaining at RAC were fundamentally no different from those made by Kaiser during the last several negotiations. He reported that both companies had spent between $2.5 and $3 million on preparations. He also admitted on the stand what the union had suspected all along—that from the very beginning the company had intended to run the plant with or without the union.

When Earl Schick took the stand, Stuart Israel got him to admit that at no time in those first weeks, after workers were sent home on the night of October 31, 1990, were they ever told that they were welcome back to work at RAC. This was a major stumble by the RAC defense team.

After the trial ended in mid-October, the lawyers on each side had until December 2 to file briefs in support of their positions. Ries refused to speculate on when his decision would be issued. Judges are not bound by time limits, and it could take months or even years for the decision. But Rich Brean was confident: "We got them. But part of it was that they just never understood what broke the law, so that a lot of the best stuff came out on direct, or badly prepared witnesses on cross. . . . The most disastrous exchange of all was between Ries and Worlledge and Schick. Either they didn't understand the issues, or wanted to appear reasonable to the judge, or just lied to us during negotiations. They killed their own case. He was fishing to find out the answers and we were just nailing their coffin shut."

In early October the union lost its appeal to Judge Charles Haden to dismiss the RICO suit. Haden was the same judge who had issued the injunction against picketing early in the lockout. Stuart Israel, who was the primary attorney on that case, suggests, "My guess is that the judge thought it would have a salutary and calming effect to have this suit pending, just to give people some discipline. It would perhaps . . . calm down those individuals named . . . or it would at least put a chilling effect on others from doing anything if they figured they could be a defendant in a civil suit."

During the last week in October, a three hundred-vehicle caravan arrived at Fort Unity, bringing more than $100,000 in donations, food, and clothing from Steelworkers and Auto Workers in Michigan and Ohio. Close to five thousand people were on hand for a Saturday afternoon rally. Locked-out worker Jeff Hill described the crowd, "You see everything here from ball coaches to preachers. There's people of all ages. This isn't a bunch of radicals. You've got good people here and they're sticking together."

The members of Local 5668 and their families were facing the one-year anniversary of being locked out. The NLRB case offered hope, as did the OSHA inspection. Compared to other recent labor disputes, to have had a wall-to-wall inspection and an NLRB trial during the first year was a major accomplishment. The end-users campaign also showed promise, with Stroh's agreeing not to buy RAC metal, and the European campaign against Marc Rich was gaining momentum. The Assistance Center was providing for members and their families and the union was holding solid—with only a tiny fraction of the membership crossing the picket line. Few other coordinated campaigns were as active and as successful on so many fronts. Yet the workers were still locked out. RAC continued to ship metal and Boyle was as determined as ever to do everything he could to beat the Steelworkers.

But for many workers, the battle was no longer just about the Ravenswood Aluminum Company. The Steelworkers had already risked too much. As Marge Flanigan put it, "I don't feel like we're fighting for ourselves. I feel like we're fighting this for every working person in America."

16

ESCALATION

Despite the breakthroughs in the summer and early fall of 1991, the lockout dragged on into its second year. Every two months Lynn Williams would ask George Becker and Jim Bowen to report to the Steelworkers executive board on the progress of the campaign. "At times I used to cringe before these damn meetings," Becker recalls. "The thought of having to think of some innovative new way that I could tell our directors that things were looking good or at least moving in the right direction. . . . At times I thought there wasn't a member on that executive board, not one besides Lynn Williams, who was rock solid in his belief that we would eventually win . . . that thought we had the chance of a snowball in hell of winning this. After one particularly depressing board meeting, a friendly director confided with me . . . 'George, how long can you take this kind of beating? . . . why don't you just stand up and tell them it's a loser and negotiate the best deal you can.'"

But Becker was no less committed to winning than Boyle or Rich. This fight had become a personal one for him, and Becker dug in his heels. When faced with criticism or doubt, recounts Becker,

> my response to this has always been . . . to escalate our activities. . . . I used to call them escalation meetings. The last thing that I wanted that company, Emmett Boyle [and Marc Rich and Willy Strothotte], to think of before he went to bed at night, Monday, Tuesday, Wednesday, Thursday, Friday, Saturday, and Sunday, is all the problems and difficulties we caused them that day. And the first thing I wanted them to think of when they woke up is, oh, Christ, I've got to go out and face them sons of bitches again. . . . We had to get them thinking about the Steelworkers continually, every day . . . if we let an hour go by that our name didn't cross their minds for some reason or another, then we were failing.

Despite moments of doubt, Lynn Williams and George Becker were determined to do everything possible, regardless of the risk and expense, to win a settlement at Ravenswood. For Williams, there simply was no alternative to pulling out all the stops. Starting with the USX lockout in 1986, the USWA president had concluded that, in the current bargaining climate, "the corporate campaign was the only way to think about it."

> The labor movement was in . . . a sort of peak of devastation in the 1980s. We'd been hammered and lost all these members, struggled with concessions. And I thought what we really ought to be doing is . . . [get] a message out to those elements of corporate America that were viewing the eighties as an opportunity to do us in. "Try if you will, it's a mistake. Your chances of winning are not great and your chances of having one terrible row are a hundred percent. Labor isn't going to sit around and . . . let our people be destroyed in this way. We're going to take you on in every way we can, and that's an enormously destructive activity. . . . It will be a major row that the union may very well win, and in any event will be extraordinarily expensive for you."

A primary part of the escalation strategy was to continue the pressure on Rich in Switzerland and the NMB Postbank in the Netherlands. After the union delegation returned home to Ravenswood in June, Joe Lang and the Swiss labor movement continued the campaign to get Rich ousted from his safe haven in Zug. A few days after the Steelworkers left Switzerland, the Christian Democratic People's Party of the Canton of Zug, a conservative political party closely identified with the church and Marc Rich, called on Rich to intervene in the Ravenswood dispute to protect the reputation of Zug as a financial center. Quite a coup for Lang and the Swiss metalworkers, this represented the first time the Christian conservatives had publicly aligned themselves with the Green-Socialist Alternative Party (SGA).

That same week SGA representatives to the Zug parliament filed an official "interpellation," or parliamentary question, calling on the cantonal government to investigate Rich and Strothotte's involvement in the Ravenswood dispute and the impact of the lockout on Zug's international reputation. In Geneva, the International Metalworkers Federation (IMF), led by General Secretary Marcello Malentacchi, joined the fray. By the end of July, Malentacchi had spearheaded a campaign to lobby members of the Swiss national government "to see if they want to have their country's name damaged by the behavior of Rich." At that time, Switzerland was applying to join the European Community. At the behest of the IMF, Carol Tongue, a prominent British member of the European Parliament, asked that body to request full information about the relationship between Rich and Ravenswood, "in view of the need for Switzerland to accept European norms of legal transparency."

In the Netherlands, the unions continued their pressure on NMB Postbank. Emmett Boyle's NMB-financed buyout of Strothotte and Bradley was due to be completed by July 15, 1991. But July 15 came and went without the deal being consummated and, a few days later, the NLRB issued its complaint with the specter of millions of dollars of back pay liability. As the union delegation had argued in their meeting with bank officials less than a month before, the liability made future investment by NMB Postbank in Ravenswood Aluminum quite risky. Although Boyle reported to the media that his buyout plan had "hit a snag" but was still in the works, NMB seemed to be backing away from the deal. Both the Steelworkers and the Dutch unions claimed victory.

The pressure campaign against NMB was not limited to stopping the buyout. Jim Hougan had discovered that the registered agent for Rinoman Investments, RAC's largest single shareholder, was the NMKB Trust, a subsidiary of the NMB Postbank. Just as in Switzerland, Dutch companies operate under secrecy laws that enable companies to invest in corporations on behalf of unidentified owners. When Hougan and Jan Bom from the Dutch union federation confronted the managing director of the NMKB Trust, he confirmed that Rinoman was a corporate middleman, but refused to provide the name of the ultimate owner, claiming "if I tell you who owns Rinoman, I will be fired."

Over the summer and into the fall, the Dutch unions persisted in keeping the Ravenswood story and the NMB connection in the media. They also continued to remind NMB that its $160 million investment through Rinoman in Ravenswood Aluminum was a substantial risk, in the context of the NLRB complaint and the back pay liability that exponentially increased each day that the seventeen hundred workers continued to be illegally locked out.

Part of Becker's escalation strategy was to expand the international campaign to every country in the world where Marc Rich was active. Over the summer, the union had made contacts with metalworker unions throughout Western and Eastern Europe, the Caribbean, and Latin America. From these contacts and Hougan's continued research, they learned about several major Marc Rich ventures in the works: an aluminum smelter in Venezuela; a luxury hotel in Bucharest, Romania; the National Aluminum Company in Czechoslovakia (now Slovakia); and oil trading in Finland, the Soviet Union, Iraq, and South Africa. The strategy team was determined to pursue all these leads.

The union researchers had assumed that, since Rich had been described as America's most wanted white-collar criminal, either the FBI or another division of the Justice Department would have produced a

"Wanted" poster years ago, when Rich first fled the country. To their surprise, after repeated calls to the FBI, the U.S. Marshal's Office, and Interpol, they were unable to track down any kind of flyer or "Wanted" poster about Marc Rich. That same fall Donald Bartlett and James Steele, reporters from the *Philadelphia Inquirer* researching Rich's role in the Ravenswood lockout as part of their prize-winning series "America: What Went Wrong," had similar luck when they called the FBI. When asked about the existence of a Marc Rich "Wanted" poster, an FBI expert on international fugitives replied, "I know the name. It's right on top of my head. It's not coming to me what he's been involved in and where he's wanted."

Uehlein suggested that the union should print its own "Wanted" poster. Featuring a sinister picture of Rich, the leaflet read, "Marc Rich, head of the Marc Rich Group companies, is wanted by the U.S. government, which is offering a $750,000 reward for his arrest." It detailed Rich's criminal record and his links to the Ravenswood struggle. The leaflet was translated into eight languages in preparation for the international campaign.

The first opportunity to use the "Wanted" poster in Europe came in early October. The union had discovered that each fall the most influential metal traders in the world gathered for a black-tie banquet at the London Metals Exchange (LME). Although Rich himself could not attend because he would risk arrest and extradition to the United States, the union expected many representatives of the Marc Rich Group. After all, Rich alone controlled 40 percent of the LME's aluminum trading and his London operation was one of his primary offices. The LME event would therefore be a gathering of Marc Rich's most powerful surrogates, competitors, and peers, and therefore the perfect place for the union to publicly press its campaign.

The union delegation for this trip included Hougan, Uehlein, Doug Bashaw, Local 5668 picket captain, and Jerry Schoonover, pot room safety committeeman. While Uehlein made last-minute travel arrangements for the union delegation back in the United States, Hougan flew ahead to London to set things up there. Hougan got the details of the banquet from Jim Regan, a reporter for *American Metal Market*. Regan also provided Hougan with an official pass to attend the dinner.

Shortly before the rest of the delegation arrived from West Virginia, Hougan returned to the hotel to find a strange man in his room. The man claimed to be a window washer hired by the hotel and abruptly fled. Hougan went immediately to the front desk to inquire and, as he expected, the hotel had no idea who the man was or what he was doing in the hotel. Back in his room, Hougan found that his briefcase had been opened and his papers rifled, but nothing was missing. He suspected that this was no ordinary burglary attempt.

The rest of the delegation joined Hougan on October 8, 1991, in time for the Metals Exchange gathering. Armed with his pass and dressed in a rented tuxedo, Hougan left piles of "Wanted" posters inside the luxury hotel. He also circulated around cocktail parties, mixing with guests and gathering information. Outside, Uehlein, Bashaw, and Schoonover, along with British union leader Len Powell from the Iron Steel Trades Federation, handed Marc Rich "Wanted" leaflets to the metal traders as they stepped out of their chauffeured limousines.

To their amazement, the Ravenswood delegation was received very well. They had already received permission to leaflet from the local authorities. Even hotel security, who rushed out when they first appeared, just politely cautioned them not to block the entrances. As Jerry Schoonover describes the scene, it was quite an experience for the local union delegation.

> We went to the hotel where the rich and the famous was having the party. Everyone was in black tux, all the men. I'd say the cheapest gown that any of the ladies had on was ten to fifteen thousand dollars, at least. . . . And, what really surprised me, here are two guys from West Virginia and Joe Uehlein and another labor man . . . we was standing there, handing these handbills out, with a big picture of Marc Rich on it. And everybody that went in, except two or three, accepted the handbill. Some of them talked about them. Some talked with us. Some of them went in and then came back out. I had a man and woman come back out and ask me about Pinky Green. They asked me if Marc Rich was the father of Catherine's child [movie star Catherine Oxenberg]. . . .They're asking all these things. But you could distinguish between which ones were Marc Rich's competition and which was for Marc Rich.

While they were leafleting, the group got their first taste of Rich's ruthlessness. Early on, a London cab pulled up alongside them with two tuxedo-clad men in back. One man jumped out and quickly asked Uehlein for two leaflets, jumped back into the cab, and drove away. Not long after, a large black BMW with darkened windows pulled up. As four large men stepped out, Uehlein went over to hand them a leaflet.

> The back door opened first and I handed the guy a leaflet. He wouldn't take the leaflet, stared at me with a very cold and serious look and said, "I

know what this is about. This is about Ravenswood. You'd better stop or you're going to get hurt." By this time all four occupants of the vehicle were out of the car and standing on the curb. They were impeccably dressed—very stylish, suit or sport coat with trousers and tie. I told them that we had come a long way to do this leafleting and we weren't going to stop. He repeated, "You'd better stop or you're going to get hurt." Pointing to the photo of Marc Rich on the leaflet, he said, "I know these people, and if you don't stop you're going to get hurt. You don't know who you're up against." Finally he said, "It sure would be a shame if you ended up at the bottom of that lake tonight," pointing to a lake directly across from the Grosvenor House hotel. Then they got in the car and left.

Joe was extremely shaken by the incident, knowing what he knew about Rich. Still, they continued to leaflet. Later two London bobbies showed up at the request of the hotel manager. They asked the workers several questions and told them they could continue to leaflet as long as they did not approach people. Instead, they were instructed to stand against the building holding up the leaflets, waiting for the guests to approach them. The union delegation complied and the guests continued to take the flyers. While the police were still standing there, the BMW returned. This time only one of the men got out; he passed Uehlein without stopping. As he went by he turned his head and said, "I warned you," and returned to the car. A little later the BMW reappeared. The four men were now dressed in tuxedos and went directly into the banquet without commenting to the union delegation.

Unaware of what was happening outside, Hougan was busy inside the banquet hall going from reception to reception placing small stacks of "Wanted" posters in each room. He also had several discussions with guests at the event, pumping them for further information on Rich and his associates.

The next afternoon the union delegation leafleted the London offices of the Marc Rich Group. According to Uehlein, within minutes of their arrival, the office went into a frenzy of activity.

A lady came out, very upset, asking who we were and what were we doing. She asked for some leaflets and I pointed out our identification (USWA) on the back of the leaflet. Next, a cab pulled up and the driver rushed in and the lady gave him a leaflet and he returned to his cab and rushed off again. We could see all of this activity through the glass walls leading into the foyer of the building. Then two guards flanked us, one on each side, while we leafleted. When we stopped leafleting and left, they followed us. We purposely did not return to the hotel, but rather walked around for about forty-five minutes until we thought they gave up, and then we went to the hotel.

When the group returned to the hotel, they discovered that Jerry's and Doug's rooms had been ransacked and Doug's camera was missing. This time the union filed a police report. Throughout the next day, they were followed by security guards carrying hand-held radios, the same men they had seen outside Rich's headquarters the day before.

While they were leafleting outside the Metals Exchange, one of the guests had told Jerry Schoonover that Emmett Boyle would be the keynote speaker at the *Metals Week* aluminum industry convention a few weeks later in Vancouver, British Columbia. It was scheduled for October 29–31, 1991, the one-year anniversary of the lockout. On the spur of the moment, the strategy team decided that Joe Chapman, Rich Yeselson, and Local 5668 members Dewey Taylor, Joe Johns, and Ed Lasko would go to Vancouver to leaflet the convention and hold a rally and press conference.

For this event the union designed a new leaflet showing Marc Rich as a puppeteer pulling the strings on a puppet labeled Boyle/RAC, who in turn was shown trampling workers holding up the state of West Virginia. Beginning "R. Emmett Boyle and Marc Rich: International Ties manipulate RAC—Destroy a peaceful West Virginia town—Victimize West Virginia's working families," the leaflet charged that Marc Rich, "working though a network of corporate fronts from his hideout in Zug, pulls 'strings' connected to R. Emmett Boyle. . . . Boyle tries to cover up his links to Rich. But Boyle can't hide the ties that bind RAC to Rich."

The group had decided against registering for the conference, to avoid exposing themselves to the sponsors or Boyle. Instead they would sneak into the conference unannounced. Early on the morning of October 29, they gathered, along with about a dozen Canadian trade unionists, outside the Hyatt Regency. As the only ones wearing suits and ties, Joe Chapman and Dewey Taylor went in ahead to case out the situation inside. They made it as far as the main hall on the third floor, where the speeches would be made, when their luck ran out. They ran right into RAC spokesperson Debbie Boger. Taylor and Chapman hurried off to find the others and warn them that they had been spotted, but they were able to continue leafleting for another hour without any interference. Rich Yeselson recalls, "Our package consisted of the Boyle-Rich [puppet] leaflet and the Rich 'Wanted' poster stapled together. Between us, we must have distributed two hundred or three hundred leaflets, obviously reaching a large percentage of the attendees. Quite a few people that I leafleted mentioned the London action, and one guy with a Shearson, Lehman ID, said that 'a lot of us have been screwed by Marc Rich.'"

When the union team got to the third floor they continued leafleting in the main hallway. Ed Lasko spotted Emmett Boyle. Lasko, a bricklayer in the plant, had known Boyle since Boyle started at RAC as an engineer. Lasko's daughter had been the babysitter for Boyle's children before Boyle and his family left for California. Lasko describes the incident:

When he seen me, he turned white. Because, see, he knew me personally. He just turned gray. So he left. He had to get Sugar-booger [Deborah Boger]. She was with him and one of his security guards was with him. The head of security for Ravenswood, he was there. And Worlledge and their secretaries and Mancini and all of them. Well, I had about six or eight pamphlets left. So, she [one of the secretaries] came over and she said, "Can I have one?" And I said, "Yeah, you can have it." . . . I said, "You can have them all if you want them." . . . So she just grabbed them out of my hand and she went over there. Someone notified security. This man come up to me and he said, "Are you with the metals conference?" I said, "I'm not." He said, "Well, I'll have to take you out of here." I said, "Oh, that's no problem."

Metals Week security guards quickly pushed the workers out of the room. Rich Yeselson was confronted by the magazine's publisher. Shoving him across the floor, the publisher told Yeselson that he "ran the conference and would not have it disrupted." Yeselson tried to convince him of the news value of the story "that was happening right under his nose" and asked for thirty seconds to explain the significance of the Rich/RAC/Boyle connection. The publisher gave Yeselson his thirty seconds but, after hearing a rapid-fire account, said he still wasn't interested and kicked Yeselson out.

Although security removed the union delegation from the premises before Boyle started his speech, and the Vancouver media provided only minimal coverage to the leafleting and rally that followed, Boyle definitely felt their presence. He was being forced to think about the Steelworkers campaign "first thing when he got up in the morning and last thing before he went to bed at night, Monday, Tuesday, Wednesday, Thursday . . . " even when he was three thousand miles away giving a keynote speech in British Columbia.

Given the secrecy that shrouded Marc Rich and his operations, a major part of the pressure campaign was to get national and international media coverage of Rich and his connections to Boyle and Ravenswood Aluminum. That fall there were several major breakthroughs, as feature-length articles about the Ravenswood campaign were published in the *Philadelphia Inquirer, Business Week,* and *The Progressive* magazine. For the first time since he fled to Switzerland, Marc Rich's dealings were under the spotlight of the mainstream media. The articles told how a notorious international fugitive was able both to sell copper to the U.S. Mint and trade oil with Saddam Hussein while he and his associates wreaked havoc on the lives of West Virginia workers. It was just the kind of media attention that Marc Rich and his lawyer, Leonard Garment, dreaded most.

Perhaps the most damaging news story about Marc Rich during this period was printed in the October 18, 1991, issue of *Suera,* a Finnish news magazine. In two in-depth articles, entitled "The Police Looking for Marc Rich—Man Who Procures Oil for South Africa Is a NESTE Customer" and "American Union Leader Warns Reputation of NESTE and Finland in Danger," *Suera* brought to light a corruption scandal involving the Finnish national oil company, NESTE, and oil and weapons trading deals with Marc Rich, South Africa, Iran, and the Soviet Union. The article also told an interesting story about a foiled attempt to arrest Marc Rich at a Finnish airport.

> On Thursday, September 19, the central criminal police received from Interpol an urgent message. On that same or following day, on a private airplane there would be arriving in Finland a Belgian-Swiss businessman, Marc Rich: arrest him for extradition. Behind the request were the U.S. federal police, the FBI, and the U.S. Marshal Service, which looks for wanted persons. Rich was looked for through the weekend at the airports, but he never showed up.

The authors of the article suspected that Rich's plane turned around because of a leak in the State Department. The FBI sent an urgent request through Interpol that Rich be arrested by Finnish police that September, but someone warned Marc Rich.

The *Suera* article and others like it threatened Rich's lucrative deals with both NESTE and the Soviet Union. But it was in Czechoslovakia that November that the Steelworkers had their first success in actually thwarting Rich in his deal making. Earlier that fall, Jerry Fernandez, USWA international affairs director, had learned from the Czechoslovakian metalworkers union that Rich was about to purchase the Slovakian National Aluminum Company. With the fall of the Communist government, previously nationalized industries were up for sale and metals traders and international financiers such as Marc Rich were poised to make billions of dollars in profit by purchasing entire industries at bargain prices.

Told that the Rich deal might go through within the month, the Steelworkers made contact with the Czech embassy in Washington, providing embassy staff with a thick information packet on Marc Rich and the Ravenswood campaign. The embassy passed on the union's concerns to the Czech government. Knowing that the sale was imminent, on November 21, 1991, Lynn Williams sent an overnight letter to Czechoslovakia's new president, Vaclav Havel. The letter briefly described Rich's unsavory record throughout the world and questioned whether the "progressive agenda" set by Havel would be furthered by his country's "entering into partnership with a criminal who represents the worst of Western capitalism."

On December 4, the Czechoslovakian embassy notified Joe Uehlein that, as a result of the union communication with the embassy and the letter to Havel, Rich's purchase of the Slovakian National Aluminum Company was on hold and under review. The union campaign was no longer costing Rich only his privacy and his peace of mind. With this action the Steelworkers had proven that they could also interfere with Rich's ability to do what he did best—trade and invest behind closed doors.

17

KEEPING UP THE PRESSURE

As the lockout in Ravenswood entered its second year, the Steelworkers and their families waited. They were waiting for Judge Ries and the NLRB to rule on the unfair labor practices. They were waiting for OSHA to make its determination. And they were waiting for the end-users campaign to convince companies to stop buying RAC aluminum. But for Paula Clemmer, a salaried worker in the plant, the wait was too much. She wrote in a letter to the editor of the *Jackson Star News*:

> During the past year I was locked in while my steelworker was locked out. Since I live with a union member people I have known and worked with for over 24 years began to avoid me. Friends have brought their sons and daughters in the plant to work as scabs, stealing Steelworkers' jobs. I received a phone call at home telling me, "If this bullshit with the Union doesn't stop your house will burn." Apparently the caller thought he was calling a union home, not a "mixed" home. For the first time in 24 and a half years I have been accused of carrying information out to the Union. I have always taken the responsibility of my job seriously, and these accusations have caused scars that will never heal. It was a hard decision to leave. I love many of the people I work with and could depend on them until the bitter end. I fear it is near. Someone will be killed before this is over if it isn't stopped soon.

Clemmer went on to lament why the conflict had gone on so long. In memory of her late father, "a union man all his life," and "for the love of many locked-out employees at Ravenswood," she was "taking the first real stand in the 42 years I have lived." She concluded by addressing Boyle directly: "Emmett, I'm so disappointed in you. I thought you were a better man. Please, for the sake of everyone, put an end to this. I always felt I would retire in Ravenswood; now, as I prepare to leave I never want to see it again."

Now the locked-out workers faced their second holiday season without a RAC paycheck. The Women's Support Group worked hard to prepare for Thanksgiving and Christmas parties that would lift everyone's spirits, especially the children's. Busy as they were, Marge Flanigan and the other leaders began to worry that the women too were beginning to be worn down by the waiting. If they lost their resolve, the men would soon follow. They decided the time had come to do something just for their members. One day, as they were sorting through eight busloads of donated clothing in preparation for the holidays, one of the women suggested that they have a "before and after the lockout fashion show."

While Marge acted as narrator, one by one the women paraded in front of the group. As Marge recalls, "Rose Cass came out in a beautiful suit and all professional-looking, and then a year later Rose is the bag lady and all her customers went away and she's forced now to live under the bridge. . . . Jody [Dean] wore the pink evening gown. She was just our little socialite, she was always fluttering around. Then she came as a cat lady, all in black with a spray can."

Another of the women, Charlotte Cobb, came first in a housecoat and then in full riot gear, while Karen Hughes, the one who had caught the goon guards in her gun sights while hunting on her property, came out first with a badminton racket and then as a goon guard. The show was incredibly funny and entertaining, and the women laughed harder than they had in months. But the transformations they presented were not so far from the truth. The lockout was changing them, changing everyone.

Ravenswood Aluminum kept up its war in the courts and in the press. Company lawyers continued to litigate on the "gag order" barring the union from using information discovered as part of Boyle's divorce. At the end of November Judge Haden once again denied the Steelworkers' motion to dismiss the RICO charges. The same week, Haden sentenced locked-out worker Robert Buck II to thirty-three months in federal prison for the March pipe-bombing incident. In return for testifying against Buck, Gerald Church received a reduced sentence of one year under house arrest.

As the holiday season approached, eight clergy, representing the major faiths in West Virginia, issued an open letter featured in newspapers in Ravenswood, Parkersburg, and Charleston the Sunday before Christmas. Their letter began, "Christmas 1991 is almost upon us with its bright promise of peace on earth to men and women of good will. Yet, here in our midst, in the town of Ravenswood, nestled close to the Ohio River, peace is an empty word. The scene is tense and potentially dangerous. Families, neighbors, and church communities are divided. To us, representing Church leadership in West Virginia, this seems to be the appropriate time, under the guidance of the Holy Spirit, to share our thoughts plainly and publicly." They went on to plead with both labor and management to "look deep into your heart" to try to resolve the situation at RAC. The letter concluded, "If we can help you bring together the jagged pieces of your broken conversation or can assist you in another way, please call us." Even the clergy, who to date had been strikingly silent about the fight in Ravenswood, knew that something would have to happen soon.

Throughout late November and early December there had been rumbling about the upcoming OSHA decision. On November 19 RAC issued a press release saying that OSHA might find as many as 275 violations, but that they would be relatively minor, with many reflecting the changing standards at OSHA rather than any serious problems in the plant. But when the decision was released on December 21, 1991, RAC was hit hard. OSHA issued 231 safety and health citations with penalties totaling $604,500. The company was cited for two "willful" violations, including "employee exposure to unsecured aluminum coils on tractor trailer beds and failure to record all occupational injuries and illnesses."

The second charge about falsification of records was exactly the claim that Bill Doyle had been making all along. In January 1991 he had made these allegations to the press, and RAC promptly sued him for defamation and libel. A defiant Doyle reacted, "In this report they get cited for failure to report. That vindicates me." George Becker went farther, "Union workers died at Ravenswood, scabs were injured, and the public was treated to a pack of lies about how good conditions are at RAC. This management ought to be held accountable for their illegal actions." One hundred and seventy-six of the remaining violations were termed "serious," with fines totaling $440,000, including "hazards related to open-side floors, personal protective equipment, exits, slings, preventive maintenance, electrical systems, and exposure to hazardous materials."

RAC responded predictably: "We have consistently maintained that we do not believe any of the items OSHA has mentioned during the wall-to-wall inspection or the prior complaint inspections have resulted in any injury to any RAC employee, subcontractor, or visitors to the RAC facilities." Bill Doyle was furious. This was the largest fine ever brought against a U.S. aluminum company, with clear evidence of death and serious injury that the company was covering up. "Just look at the citations and fines proposed," Doyle said. "They speak volumes about how little management has done for the safety of the workers. These huge fines aren't for minor 'parking-ticket' violations. These are very, very serious, unsafe, life-threatening conditions RAC has been hiding."

The OSHA ruling was a tremendous victory for the Steelworkers. Not only was it a vindication of the union's allegations about health and safety, it was a publicity nightmare for the already embattled company. But as Christmas came and went in Ravenswood, the members of Local 5668 were still locked out. Victory remained elusive.

One of the challenges facing Becker and the strategy group was how to keep up the pressure on both the international and the domestic fronts. "So you'll see in July and August a whole series of memoranda [saying] we're letting Emmett Boyle off the hook, there's no pressure on Emmett, we've got to get him," recalls Joe Uehlein. "And then in the fall we started thinking, oh my god, we're not escalating overseas anymore. God, we've got to get back over there. And so then you'll see a fall offensive overseas. So there was this back and forth."

Upon returning from Europe in July, Uehlein started floating the idea of a "stakeholders" report and meeting. Since RAC was a privately-held corporation, there were no stockholders, stockholders' reports, or stockholders' meetings. Uehlein explains, "In recent years the labor movement has developed the concept of *stakeholders'* rights in a company, as opposed to those of *shareholders.* Stakeholders may include customers and suppliers to the company; small businesses that depend on the clientele generated from the company's workers; community political and religious leaders . . . and, most importantly, workers—whose incomes, self-respect, retirement prospects, and connections to the community at large as citizens—are all tied to their employment at the company."

The idea was to first issue a "Stakeholders' Report," modeled on corporate annual reports, that would document the damage RAC had levied on its stakeholders. The report would then be followed by a stakeholders' meeting. This would not be a typical rally of union members railing against RAC; in many ways the union would take a back seat to the other stakeholders, who would make their own case against the company. Joe Chapman told the press, "We'll give a report of the state of the company as we know it."

The strategy committee liked the stakeholders idea, but debated it endlessly. And although Becker encouraged it, trying to coordinate with the USWA public relations department and outside media consultants turned out to be a very slow process. The first draft of the report used a tongue-in-cheek tone that was not at all in keeping with the more dignified approach Uehlein and the strategy committee had envisioned. They lost another several months. "George was beside himself at certain points as to why we couldn't get this stakeholder report done," recalls Uehlein. "I remember just before Christmas, sitting in George's office, and we're like pulling our hair out. 'Why hasn't this thing happened? What's the problem?' George just said, 'Look, here's a date. Is it okay with you?' I said, 'Yeah, that's fine with me.' He said we're going to have the stakeholders' meeting on that date, and he sent out the word to Frank Powers, Gary Hubbard, Jim Bowen. The report better be done, because there's the meeting date."

The final stakeholders' report looked like a corporate annual report. Printed on heavy, glossy stock, its cover featured a stylized map of the world, zooming in on West Virginia and Switzerland. The report contained the traditional "Letter from the Chairman"—in this case from George Becker—as well as discussions of operations, ownership, government relations, and potential liabilities. For the first time Marc Rich's tangled empire and the union's international strategic campaign were laid out in detail. As Rich Yeselson later explained, "After seeing the stakeholders' report, even the most apathetic union worker understood the Marc Rich angle and how much the international union was doing on behalf of the membership."

The Steelworkers pulled out all the stops on organizing for the stakeholders' meeting. Advertisements inviting stakeholders were placed in seven local newspapers. On January 22, 1992, more than two thousand locked-out workers, their families, union supporters, religious leaders, small business owners, and politicians turned out for an emotional event at the Charleston Civic Center. It was a major media event covered by all the local television stations, as well as several networks.

After opening remarks by Jim Bowen, Dan Stidham, Joe Chapman, and Dewey Taylor, Bishop Bernard Schmitt from the Catholic Diocese declared: "To replace striking workers on a permanent basis is certainly against the rights of unions and the rights of individuals. . . . The role of unions in promoting the dignity of workers is important in Catholic teaching. Collective bargaining is essential to protect workers." Community leaders spoke as well, including Foodland manager Jerry Carpenter, who had done so much to support the Assistance Center.

George Becker outlined the information contained in the Stakeholders' Report and Governor Caperton posed the question, "Is it fair that one of our great communities is torn apart by a labor dispute that didn't have to happen? . . . Is it fair that seventeen hundred people, mostly older workers, are suddenly without the chance for a job? The answer is clearly no." The governor went on to challenge both sides in the dispute to meet in his office. Jim Bowen accepted the offer on the spot. Later in the day RAC refused.

The crowd went wild when a giant puppet of labor hero "Mother Jones" emerged, chastising smaller puppets of Emmett Boyle and Marc Rich. Inside Mother Jones was Tavia LaFollette, a student at Antioch College in Ohio, who had created Mother Jones with her teacher Amy Trumpeter as part of a class project for use during the UMWA's Pittston strike. LaFollette is the great-granddaughter of "Fighting Bob" LaFollette, the Wisconsin Senator and leader of the populist Progressive movement in the late nineteenth and early twentieth centuries. During the Pittston strike, Tavia developed "agit-prop" skits that were wildly popular with the miners throughout the southern coal fields.

Tavia describes her creation: "Mother Jones is fifteen feet tall. She's made out of papier-mâché and cloth. I wear a harness, almost like a flag holder, and there's a stick that goes to the top of her head, and I put the stick in the flag holder and then her dress and everything is like a tent, tepee, around me. So I'm actually underneath, inside of her dress and speaking from within. And I need two other people to operate the hands. So they do the gestures of her. And each hand is on a long stick." LaFollette recruited Steelworkers to play Emmett Boyle and Marc Rich. Instead of full-size puppets, the Boyle and Rich figures were more "like a distorted head that fits on top of your head so you look kind of goofy." The scene made great theater.

It was Joe Uehlein who came up with the idea of using Mother Jones in the Ravenswood campaign. A longtime guitar player and singer of labor music, Uehlein is president of the Labor Heritage Foundation (LHF), a nonprofit organization that promotes the use of music and culture in the labor movement. The LHF holds an annual conference in Washington each year, and Amy Trumpeter had brought Mother Jones and talked about her work at Pittston. Given Mother Jones's West Virginia roots, the puppet was a natural for the Ravenswood struggle. After Uehlein recruited LaFollette for the Steelworker campaign, he suggested she make the smaller Rich and Boyle puppets as well.

While less theatrical, the first public discussion of a growing environmental campaign was also held at the stakeholders' meeting. Unlike the end-users campaign and the international campaign against Marc Rich, which depended on publicity for their effectiveness, the environmental work to date had been behind the scenes. For almost a year the strategy group had been making steady progress, examining environmental issues and making contacts in the environmental community.

In May 1991, Joe Uehlein had contacted John O'Connor, director of the Boston-based National Toxics Campaign Fund. Through O'Connor the Steelworkers were put in touch with Robert Ginsberg, a Chicago-based environmental consultant, who conducted the initial research on RAC. His June 1991 report outlined several serious issues.

The first concerned something called "spent potliner." Over time the carbon lining of the pots used to smelt aluminum become saturated with cryolite and aluminum fluoride, causing the carbon lining to crack and making replacement necessary. The spent potliner removed from a single pot weighs between thirteen and fifty tons. For years Kaiser simply dumped this waste on-site near the plant. In addition to containing heavy metals and other toxic substances, when the spent potliner, or cathode waste, as it is called, gets wet, it produces cyanide. During the 1960s and 1970s cathode waste contaminated the groundwater at the plant with cyanide. When "they were dumping it outside, all the pigeons were dying alongside the road to the river," recalls Bill Doyle. "And I told the industrial hygienist, 'I wonder if that's cyanide water,' and he said, 'Oh, yeah, you can bet on it.'"

When Kaiser sold the Ravenswood facility to RAC, Kaiser retained the liability on what Bill Doyle estimated was one-half million tons of spent potliner. Doyle was concerned about the transportation of this material on railroad cars and barges and the environmental disaster that would occur in the event of an accident. "If one of those barges upsets in the Ohio River, you're going to kill all the fish between here and the Mississippi River." While Kaiser had retained the liability on the solid cathode waste, RAC was responsible for the contaminated ground water routinely released into the Ohio River. The union needed to understand more about the amount of cyanide being released, as well as the permit process.

On July 16, 1991, the Steelworkers contracted with Ben Ross of Disposal Safety, Inc., to conduct a study of RAC's "spray fields." Since 1972 the company had been disposing of the water that originated in the hot rolling mills—which contained oil, grease, and solvents—by spraying it in fields near the plant. According to Disposal Safety, "The purpose of the spray fields is to biologically degrade the oil and grease into the water and to fix the lead to the soil." Based on his research, Ross was convinced that, because the water was contaminated by solvents, the spray fields were in violation of the Clean Water Act. In late August 1991, after reviewing Ross's initial research, the West Virginia Department of Natural Resources (DNR) initiated an inspection. Just like the inspectors from OSHA, they were refused entry. Debbie Boger moved swiftly to control the damage and an agreement was quickly made to allow the investigation.

The union continued working on these environmental issues quietly, behind the scenes. A copy of Ross's report, released in October, was sent to the NMB Postbank in Amsterdam and to the DNR. Ross, his assistant Steve Amter, Doyle, Yeselson, and members of the West Virginia Environmental Coalition met with DNR officials in Charleston on October 25, 1991.

On January 25, just a few days after the stakeholders' meeting, the DNR demanded documentation from RAC on its waste disposal procedures. Ben Ross was quoted in newspaper reports: "Ravenswood Aluminum Corp. has been allowed to systematically ignore the requirements placed on other factories. RAC represents a case study in environmental non-enforcement." DNR officials were hard-nosed. Spokesperson Brian Farkas reported that the spray field technology was obsolete. "They use an oil and water mixture to cool drums in the hot rolling mills. The company says the drums are so hot that the oil is vaporized and only water is left." DNR was not certain the oil was gone, and requested documentation to determine what was actually sprayed in the irrigation fields. "The U.S. Environmental Protection Agency issued a permit to Ravenswood that allows the company to dispose of cyanide in the river," Farkas went on to report. Then he dropped the bombshell: "But the permit, called a National Pollutant Discharge Elimination System permit, has expired."

The news hardly inspired confidence in the already troubled company. It was another public relations nightmare, and one that garnered further support for the Steelworkers from state and national politicians. The environmental charges only increased RAC's rogue status. Action against RAC was initiated on both state and federal levels. On February 12 and 13, 1992, the West Virginia Senate Judiciary Committee held widely publicized hearings on the Ravenswood lockout. In a thinly veiled reference to Marc Rich, Chairman Jim Humphreys, D-Kanawha, began the meeting by stating, "The chairman recognizes the problems created by criminals owning West Virginia corporations." Joe Uehlein testified about Marc Rich's tangled web of connections in the aluminum industry. Jim Bowen took the web metaphor one step further, telling the committee, "The web is there and the spider is crawling around in the web, and it is poisoning the people who have given their lives to Ravenswood Aluminum."

Joe Lang traveled from Zug to testify at the hearings. The presence of a Swiss citizen and legislator made a strong impression. The *Jackson Herald* reported, "While West Virginians have only heard stories about Marc

Rich and his illegal activities from members of the Steelworkers of America, Lang can address the issues firsthand, as a citizen of Zug, Switzerland." Issuing a stinging rebuke to Rich, Lang reported, "The fact that he denies his relationship with RAC isn't even taken seriously by his local lobbyists in Zug."

Despite the mounting evidence, RAC officials continued to vehemently deny any connection to Rich and refused to attend the Senate hearing. Chairman Humphreys called the company's refusals to appear "shocking,"adding, "it is regrettable that one side adopts the position that they are above public accountability or responsibility to the government." Humphreys directed state attorneys to draft legislation that would allow the state to deny permits to companies owned or controlled by fugitives.

The hearing was also covered widely in the press. Media consultant Frank Powers reports, "We went out that night to catch a flight up to Charleston Airport. We got there, and . . . we're watching the 6:00 news. . . . The waitresses in the restaurant there said, 'Boy, look at this. Wow.' It had an enormous impact."

In many ways the West Virginia hearings were a rehearsal for upcoming Congressional hearings scheduled in Washington. By a stroke of luck, the chair of the House subcommittee dealing with "Government Information" was Bob Wise from West Virginia. Since the Workers' Memorial Day rally in April 1991, where Wise spoke and later clog-danced to the Paw-Paw band, he had continued to be a strong supporter of Ravenswood workers. Wise held hearings on December 4, February 18, and March 5. While some might have dismissed the West Virginia hearing as a relatively routine action of an isolated and small state legislature, and therefore not particularly damaging to Rich, hearings held before the Committee on Government Operations of the U.S. House of Representatives were another story. More than 250 pages of testimony and documents would be published and made part of the Congressional Record. "The purpose of today's hearing," opened Bob Wise, "is to continue to try to untangle the bureaucratic web which is somehow preventing the U.S. Government from stopping its business dealing with a firm connected to a man indicted in the largest tax evasion case in the nation's history."

The committee discovered that from 1985 to 1988 Clarendon had been prohibited from bidding on U.S. government contracts because of the company's links to Marc Rich and his illegal activities. However, when this three-year debarment ended, the government failed to conduct an adequate investigation of Clarendon, as required by law. In 1989, soon after the debarment was lifted, the U. S. Mint began to receive many complaints about its contract with Clarendon. In November 1989 the

Treasury Department launched an investigation, which the Wise Committee discovered had made stunningly little progress, while, in the meantime, the Mint continued to award contracts to Clarendon.

The second focus of the hearing was potentially even more damaging for Rich. Wise asked, "The second track is why don't we have Marc Rich back? . . . This isn't your miscreant who has fled the country for knocking over fifteen 7-Elevens and is kicking around the docks of Marseilles. This is Marc Rich operating with total impunity out of a tall office building in Switzerland. And why hasn't this been made a high priority?"

The committee lost some steam when on March 5, the second day of hearings, Clarendon announced that it would temporarily withdraw from bidding on any U.S. government contracts. But Wise went on to focus on why Rich and Pincus Green had not been apprehended. Over three days the committee heard from a dozen witnesses and reviewed another dozen statements and hundreds of pages of documentation. The hearings uncovered several troubling aspects of the effort, or, more precisely, the lack of effort, to collar Marc Rich. Although the Justice Department repeatedly tried to justify its lack of progress in tracking down and arresting Rich and Green, officials admitted that with a commitment of resources from the highest levels of the Department, the government could have brought both men to justice. As Wise discovered, the Justice Department simply "lacked the political will to apprehend these fugitives." There had been no "Wanted" poster, no published reward for their arrests, not even a listing among the "fifteen most wanted criminals" sought by the U.S. Marshals Service.

The secrecy around the Rich case particularly troubled Wise. He had asked the Justice Department for information on recent contacts with Rich's representatives, but the Department refused to provide information to either the committee or law enforcement agencies. They wrote, "The Department of Justice has had contact with representatives of Rich and Green concerning the possible terms of a plea and/or surrender. It is not appropriate to disclose the particulars of any such contacts." The Wise committee responded, "The committee does not understand how 'effective law enforcement' is implicated by identifying attorneys representing fugitives from justice who seek to negotiate a plea agreement on behalf of their clients. Even the attorney-client privilege does not generally protect the identity of the client, much less the identity of the lawyer."

The evidence uncovered by the Wise committee pointed to more than just complacency in the Justice Department's halfhearted efforts to apprehend Marc Rich. Jim Hougan argues that Rich's connections with the Justice Department went far beyond former Department official J. Bradford

Reynolds, who was now one of Rich's attorneys. "Why Justice should stonewall Congress on behalf of a fugitive is uncertain, though few would doubt that the wall was built to conceal the fact that Rich is working with Justice (and quite possibly with other agencies) on what can only be called 'special projects'" [intelligence work].

Howard Safir, former Associate Director for Operations at the U.S. Marshals Service, testified that the lack of will to aggressively pursue Marc Rich went well beyond the Justice Department. According to Safir, "if a political decision was made at the highest levels of this government that we were going to apprehend Marc Rich and Pincus Green and use all available tools . . . we would have Marc Rich and Pincus Green very quickly."

On April 30, the committee issued two blistering reports: "Coins, Contracting, and Chicanery: Treasury and Justice Departments Fail to Cooperate" and "They Went That Away: The Strange Case of Marc Rich and Pincus Green, a Rich Associate." While no further legislative action was scheduled, Marc Rich was back in the limelight.

Closer to home, the Ravenswood lockout erupted in the West Virginia legislature again in early March. At a much higher level than the previous state hearing, House Judiciary Committee Chairman Jim Rowe asked for a ten-member committee with subpoena power to investigate the Ravenswood lockout. When the probe was supported overwhelmingly by a House vote of 88 to 6, Emmett Boyle sent a five-page letter to members of the West Virginia Senate, arguing that all his facilities were union and "that neither ORALCO nor Emmett Boyle support or condone 'union busting.'" He argued that, despite the NLRB charges, the Steelworkers, not RAC, had bargained in bad faith, and enumerated how much RAC contributed to the state's economy. Despite Boyle's letter, the Senate confirmed the probe, scheduling it to begin the following month.

The Steelworkers' end-users campaign continued to bear fruit. Anheuser-Busch (Budweiser) and Stroh Brewery Company had already stopped using aluminum from RAC. Although not at the time a purchaser of RAC aluminum, the Miller Brewing Company agreed not to use can stock from Ravenswood for the duration of the lockout. The Steelworkers discovered how vulnerable beverage companies can be to even the hint of action directed at consumers. A top executive at Coke told George Becker, "We don't go through all this advertising to gain market share. We do it to keep from losing market share." Becker recalls, "In effect, she said, 'We're not going to screw with the union or sue you because that would result in negative publicity. It would take market share away. In other words, we have to deal with you and your union.'"

For three major purchasers of RAC metal to have conceded to the Steelworkers' campaign was quite a coup. Company officials could not hide their anger. "The Steelworkers say they want to come back to work and if they continue this campaign and they're successful, there won't be any jobs for anybody," Debbie Boger told the press.

The environmental campaign also picked up momentum during February and March. Jim Valenti, the USWA health and safety staffer, began looking at the environmental issues with the same kind of passion that he and Doyle had devoted to the OSHA complaint. Valenti's interest in environmental issues began with workers' safety concerns about the release of chlorine gas. Kaiser had maintained a joint labor/management chlorine committee that RAC had disbanded. Although no releases had been reported by RAC, Ravenswood workers had reported many incidents to the union. Valenti explains,

These regs really meant something in terms of safety and health in the community. This was a small community. Our members' families lived in this community, and it hit me like a ton of bricks that this company really was bad for the workers and bad for their families. It just hit me that strong. I said, "I'll be a son of a gun, you can't separate the two! . . . what good is it if we cleaned up the so-called workplace, if we just vent everything to the outside and let it go a half a mile downwind and poison our guys when they're home with their families?"

A problem with moving into the environmental area was that, although the union had experience in dealing with air quality, it had done little with clean water issues. Valenti began building bridges with environmental groups who had more experience with regulation in these areas, working closely with Norm Steenstra from the West Virginia Environmental Coalition.

The environmental arena was not without risks. From the research that had been conducted thus far, RAC obviously had serious long-term environmental problems. Despite what Boger or Boyle might have argued, the strategy committee was not interested in breaking the company. Joe Chapman and Jim Bowen were adamantly opposed to the environmental campaign because they knew that it would have to take place through a coalition of labor and environmental groups and they were concerned about how much control the Steelworkers would maintain over a coalition. The campaign also raised important ethical issues. "If they've got a big problem with toxic waste, the day they sign the contract and settle the labor dispute, that problem doesn't go away," explains

USWA safety and health director Mike Wright. "And it's not ethical to try to make it go away. If you build a coalition of environmental groups, you can't just leave them in the lurch. If you do, you'll never be able to count on their support again."

Despite these reservations at the international level, the local was very receptive to the environmental campaign. "They saw the light a hell of a lot faster than I saw it, and I was assigned to kind of lead the charge in the thing," reports Jim Valenti. "They say anything we got under terms of the environment we should release to the general public. After all, they've got the right to know. They really feel that way. And if it's something that means closing this plant . . . either fix it or close it down. That's Stidham's exact words and the committee's. That's the way, if it's going to affect our kids, the water we drink or air, close it down. Seriously. And they really believe it."

Given the commitment of the local union and Boyle's intransigence, the Steelworkers decided to move forward on the environmental campaign. They worked with environmental groups, and on March 25, 1992, a citizens group filed a legal notice of intent to file suit against RAC. Under the Clean Water Act this notice is required before the actual suit can proceed. The process also bought the Steelworkers time. Filing the suit were "a farmer, fisherman, swimmer, boater, water skier, and landowner adjacent to the river." One of the plaintiffs, Candie Good, reported, "The river is important to our community. We need to make sure it is cleaned up." Papers were filed on the release of cyanide into the river from the spent potliner, as well as on the unregulated spray fields. "I am outraged to see a plant of this size allowed to operate without environmental permits," remarked Norm Steenstra. "I applaud the initiative taken by these citizens, and we will do what we can to help them in court."

Once the intent to sue was filed, the West Virginia Department of Natural Resources was bombarded by letters from every major environmental group in the country urging them to move on the Ravenswood situation. Clean Water Action, Friends of the Earth, National Toxics Campaign Fund, National Resources Defense Council, Ohio Valley Environmental Coalition, and the West Virginia Environmental Coalition all wrote to Ed Hamrick, the director at DNR.

On April 3 the groups held press conferences in Charleston and Washington. Diane Cameron, an environmental engineer with the Washington-based Natural Resources Defense Council, argued that "RAC is using the Ohio River to dilute its toxic waste." Dianne Bady of the Ohio Valley Environmental Coalition accused RAC of "flaunting the law," threatening the health of "thousands of people." Senator Jay Rockefeller reported that he had asked both the U.S. Environmental Protection Agency and the West Virginia DNR to inspect the facilities. Less than a week later both agencies were in Ravenswood for surprise inspections of the plant.

This was exactly the kind of escalation that the strategy team had envisioned. The results were far more than most unions dream of: hearings in both the U.S. Congress and the state legislature, an enormous OSHA fine, a major environmental suit pending, a likely NLRB decision that would include more than a hundred million dollars' worth of back pay, at the same time that RAC customers were falling by the wayside. The local membership was also holding firm. Something had to break.

18

PICKET LINE AROUND THE WORLD

Starting in December 1991 the union's focus shifted back to the international campaign. The strategy team established three criteria for determining where the union should concentrate its actions. First and foremost, the target country had to have businesses or industries in which Marc Rich held or sought to hold a major interest. Second, there should be trade union allies in that country that could be mobilized. And, third, they wanted an element of surprise that could catch Rich off guard.

Using these criteria, the strategy team narrowed their list to Rich's operations in Switzerland, the Netherlands, England, Spain, France, Finland, Czechoslovakia, Bulgaria, Romania, Russia, Israel, Venezuela, Australia, and Hong Kong. The goal was to organize a "Marc Rich Tour," actions in seven countries in seven weeks, beginning in mid-January. It quickly became apparent that this kind of effort could not be run out of Washington and Pittsburgh. The union needed someone based in Europe who was familiar with the unions and governments along each stop of the "tour."

So in December 1991 they recruited Penny Schantz as the campaign's on-site European Coordinator. Schantz, an experienced trade unionist, had worked with both the Service Employees International Union (SEIU) and the Hotel and Restaurant Employees (HERE) and was the former president of the Santa Cruz Central Labor Council. She had recently moved to Paris, site of the AFL-CIO's European Office, and was looking to do freelance work with American unions. Schantz had known Uehlein for sixteen years and had met with him during his first trip to Zug in June 1991. She had years of experience running union campaigns in the United States, and she had connections with the international trade secretariats (groupings of national unions by industry, such as the metalworkers' and food workers' secretariats) and their affiliates throughout Europe. As Penny told it, her "short job description was trying to make Marc Rich's life as miserable as possible."

The first action of the seven-week marathon was set for Switzerland at the end of January. The month before, Hougan had met with Swiss union leaders, representatives from the IMF, and Joe Lang to set up actions at the Swiss Parliament in Bern and in front of Rich headquarters in Zug. In early January, Joe Uehlein met Schantz at the annual International Trade Secretariat (ITS) Conference in London. Uehlein briefed the ITS officials on the Ravenswood campaign and enlisted their support for the upcoming tour. In addition to renewed pledges of support from the IMF and the ICEF, the International Federation of Journalists agreed to provide Schantz with press contacts in each of the target countries. The International Transportation Federation offered to track Marc Rich's shipping operations. The Public Service International, which includes unions representing border guards around the world, agreed to distribute copies of the Marc Rich "Wanted" poster to all their members, asking guards to report Rich if he was spotted attempting to cross into or out of their countries.

Hougan, Uehlein, and Schantz worked hard to enlist media coverage for the European tour. Their efforts paid off when they learned that the union delegation would be joined in Switzerland by Scott Spencer, a freelance writer on assignment for *Rolling Stone* magazine and an old college friend of Jim Hougan, as well as "NBC Dateline" reporter Brian Ross and a Swiss crew filming for "ABC News." Joe Lang and the Swiss Metalworkers (SMUV) issued a series of press releases about the upcoming USWA visit, promising "two giant puppets representing Marc Rich and Mother Jones, the legendary leader from the founding days of the U.S. labor movement."

On January 27, Hougan, Uehlein, Mike Bailes, and Dewey Taylor, along with Tavia LaFollette and the Mother Jones and Marc Rich puppets, met in Washington to fly to Switzerland. To Tavia's horror, when she arrived in Dayton for her flight to Washington, with the puppets encased in their eight-foot plywood boxes, the airline refused to let her ship the puppets as baggage, even though she had gotten the dimensions for the boxes approved in advance. She had no choice but to saw the Mother Jones puppet in two and have the puppets follow the union delegation in a cargo jet, at twice the cost of a regular passenger flight. Worst of all, the puppets would arrive in Zurich the following day, too late to get to Bern in time for the first demonstration in front of the Swiss parliament.

Tavia showed up in Switzerland, she says, "puppetless. I arrived there and they wanted visuals. 'What do you mean you have no puppets?' So we got a sheet from the hotel and I painted up the 'Wanted' poster that night. I had jet lag, I hadn't slept for twenty-four hours, I'd been thinking a bit about Marc Rich, and this is like something out of a James Bond movie. I'm sitting there smoking cigarettes nervously in my room looking out the window at a car parked down the street, being like, oh, my God, he sees me painting his 'Wanted' posters."

Early the next morning, Bailes, Taylor, Uehlein, and a small delegation from the SMUV stood in front of the Parliament handing out Marc Rich "Wanted" posters to the ministers and staff entering the building for the 8:00 A.M. session. They were backed by the bed-sheet portrait of Rich that Tavia had painted the night before. After a cold hour of leafleting, the union delegation moved inside to meet with sympathetic members of Parliament, who pledged their support. At a press conference, Dewey and Mike told the large gathering of newspaper and television reporters about the damage that Marc Rich and his surrogates had inflicted on their community, turning their "American Dream into a nightmare." Joseph Fischer from SMUV charged that "the lever for solving the conflict is in the hands of Marc Rich."

Meanwhile, Tavia and Jim Hougan drove off to Zurich in a rented truck to pick up the puppets in time for another demonstration in Bern that afternoon. On the return trip, while Hougan drove furiously on unfamiliar roads, trying to get back in time, Tavia was locked in the back of the truck, struggling to unpack the puppets and piece them back together. Suddenly it dawned on Jim that for all he knew she might have been crushed by the puppets or fainted for lack of oxygen. He quickly pulled over and rushed to unlock the truck, only to have Tavia send him back to the cab and on their way. As Scott Spencer recounts, they arrived just in time:

> Tavia finally returns from Zurich with her puppets. The workers have been waiting in a little pizzeria near Parliament, and when she arrives they unload Mother Jones, who is now paraplegic, having had her spinal cord severed in Ohio. Michael goes off to look for a hardware store where he can purchase a clamp to make her whole again, while Tavia and Dewey lay her out on the cobblestone Bern street.
>
> Eventually, Tavia's persistence and Dewey and Michael's know-how get the puppet back in one piece. Tavia hoists it up, putting herself somewhere inside of it, while the guys from Ravenswood manipulate its massive hands and arms. Joe puts the somewhat smaller Marc Rich puppet over his head and shoulders, and they go back to Parliament, where a few TV cameras have been waiting for them.

With the giant Mother Jones and Marc Rich puppets parading up and down in front of the Parliament as their backdrop, the union delegation passed out thousands of "Wanted" posters. Joe and Tavia felt greatly relieved to finally have the puppets up and working the crowd. After all the red tape, the expense, and the delays, they felt vindicated to see the puppets accomplish just what they had hoped—turning a small action into a major media event.

The next day the delegation moved on to Zug, hoping for a meeting with Willy Strothotte. The day they had arrived in Switzerland, Agostino Tarabusi, SMUV president, had unexpectedly received a long letter faxed from Strothotte criticizing the Swiss union for having "fallen victim to the flagrant misinterpretations of the American Steelworkers' union" and for "sending out frivolous slander into the world" regarding Marc Rich's relationship with Ravenswood Aluminum. In the letter Strothotte continued to deny any connection between Ravenswood and Clarendon or Marc Rich. But to the delight of the Steelworkers, his letter closed, "The undersigned would be pleased to make himself available for further explanation or a personal conversation."

Tarabusi faxed back a reply, pushing for a meeting that same day. They all knew it was unlikely that Strothotte would meet with them, but they saw in his letter an opportunity to signal to the media in Switzerland, and to management back home in Ravenswood, that their European offensive was beginning to bear fruit. As they hoped, the Swiss newspapers jumped on the story of a possible breakthrough. The headline in the January 29 issue of the Bern newspaper *Das Bund* read, "Negotiations with Rich?" and the article went on to almost directly quote from the union press release: "In the face of a Europe-wide campaign, RAC owner Willy Strothotte signaled a willingness to talk with SMUV." That same day, the *Berner Tagwacht* heralded that the labor conflict had "entered a new phase."

After this successful day in Bern, the team was caught up in the excitement. They were ready to return to Marc Rich's town and demand a meeting. As they set off for Zug, Dewey told Scott Spencer, "I want to talk to Marc Rich. I want to look him in the eye and ask him, 'Mr. Rich, what are you going to do to get our jobs back?'"

They were met at the Zug train station by their old friend Joe Lang. As they sat in a cafe across from Marc Rich's blue-glass headquarters, they watched a very different crowd assemble. This time a much larger contingent of Swiss reporters was joined by American media, including "ABC News" and "NBC Dateline." Spencer describes quite a spectacle:

It's time to bring the puppets to the front of Marc Rich headquarters. Tavia is inside Mother Jones, with Dewey and Michael helping, and Joe Uehlein is inside Marc Rich. The director of police [a member of Joe

Lang's Alternative Green Party] fades away, since it wouldn't do for him to be there if there's an infraction of the law.

"Hi!" Tavia shouts from within her puppet, and her voice comes out immense and throaty, like an elegant Janis Joplin. "I'm Mother Jones, and my address is like my shoes, it is with me wherever I go. Come out here, Marc Rich, and face the people. You, Marc Rich, who picked the pockets of the government that tried to put you in jail."

"Yes, I did," says Uehlein, inside the Rich puppet.

The blue-glass windows are jammed with Marc Rich employees looking out onto the street. They are waving; they seem amused. The few pedestrians who come by seem somewhat amused but wary, too; they don't have the shatterproof glass to protect them if this gets any weirder.

As Mother Jones chants, "Pray for the dead, and fight like hell for the living," Tavia and Joe move closer and closer to the building's entrance. There are photographers everywhere.

Just as they had in June, the union delegation, followed by Brian Ross from "NBC Dateline," tried to enter the building. But this time the door was locked. After a long wait, a security guard came to the door and asked if there were a "Taylor and Bailes" in the group. When Dewey and Mike identified themselves, the guard said they could enter. But when Uehlein got out from under the Marc Rich puppet and tried to join them, a voice boomed out, "Uehlein, nein!" Brian Ross, too, was denied entry, so Dewey and Mike walked in, alone and uneasy. Dewey recalls:

So we went in, just to the lobby area. And we were confronted there by a tall guy that spoke very broken English. A couple of other security guards [made clear] this is as far as anybody goes. . . . And gave me the letter and it was in an envelope and I didn't look at it. But I said, "Well, we've come half way around the world and this is the second trip. And we have a labor struggle going on back in the States that we think that Marc Rich can help us with. And really we think that it could be helpful to us if we can get to talk to him." He kept saying, "Not possible." We kept insisting. . . . But after a few minutes the guy said, "You should go now and I wish you well." . . . And almost like it was something that wasn't intentional at all, he walked over to the flower arrangement and broke two lilies off the flower arrangement and he came back and handed Mike one. And he handed me one and he said, "You should go now," kind of sternly.

The tone had become menacing enough that Dewey and Mike decided they should leave. Just as they were coming out, with their yellow lilies and the letter in hand, they met up with two very familiar-looking, well-dressed men trying to get in. They were Marc Rich's attorneys, Leonard Garment and William Bradford Reynolds. They had come to meet with their client, only to find themselves locked out of the building along with everyone else. As Uehlein tells it, they had also walked right onto the "NBC Dateline" set.

Garment and William Bradford Reynolds come up, unbeknownst to us. And they didn't know what we were, what was going on. Leonard Garment literally walks up next to me, because I'm right at the front door and he says to the guy, "I'm Leonard Garment and I'm here to see Marc Rich," and the guy says, "The building's closed." And he says, "GR-RRGGRR." He gets all huffy. He says, "I'm Leonard Garment." I turned to him, I say, "Hey, Leonard, you're locked out." And he looks at me and I got one of these like out-of-context stares. Who? What? Where? And then it dawned on him. It struck him. "Oh, god, it's these guys. Yeah, right." So we knew Rich was there. Brian Ross . . . turns to Reynolds with microphone, cameras, and he says, "Hey, Brad, what's it like to be working for the government one day and the most wanted fugitive the next?"

After Garment and Reynolds went into the building, the crowd gathered around Dewey and Mike to find out what had happened inside. Dewey read the letter from Strothotte. It was short and to the point. Neither Rich nor Clarendon were in any way connected to Ravenswood. Strothotte was simply a RAC shareholder and a member of the board of directors and therefore not in a position "to receive a Union delegation to discuss or negotiate any aspect of the current labor dispute."

As the rally wound down and the crowd of reporters began to pack up their gear, Taylor and Bailes took one of the Marc Rich "Wanted" posters and, as a parting gesture, laid it against the door. Then they placed the two lilies on top of the poster and left for the hotel. That night the group heatedly debated the symbolism of the lilies. No one could decide whether they represented a conscious death threat, a friendly gesture, or just an impromptu prop to help guide them out of the building.

The next day they had planned to drive up the mountain to the World Economic Forum in Davos. They had learned from a Swiss reporter at the Bern press conference that it was an annual gathering, high up in the Swiss Alps, of influential business leaders and government ministers from all over the world. Rich had attended the Forum for the previous four years and was certainly expected this year, along with CEOs from the world's most powerful corporations and a diverse sample of political luminaries, ranging from U.S. Senator Bill Bradley to Chinese Premier Li Peng. Uehlein and the group jumped at the chance to try to stage another meeting with Rich or at least distribute massive numbers of "Wanted" posters and the newly printed Stakeholders' Report.

The logistics would not be easy. They were scheduled to fly out of Zurich that same night, and driving a van large enough to hold the puppets would be difficult on the steep mountain roads in the middle of winter. They were finally ready to go when Joe Uehlein received a call from George Becker, canceling the trip to Davos and urging them to come right home. Uehlein argued hard that they were going to miss a "beautiful opportunity" to embarrass Rich, but, according to Joe, Becker held firm, telling them "the roads are bad, just get home as soon as you can."

Puzzled by Becker's decision, they had no choice but to comply. Word had already gotten out to the media that they were planning the trip, so they used the Strothotte letter as an out. Telling the press they would "rather negotiate than demonstrate," they claimed that they were calling off the planned action in Davos as a "goodwill" gesture toward Strothotte and Rich.

George Becker kept the true reason for calling off the Davos trip to himself. Becker had received a message that if the group went ahead with the trip to Davos, they wouldn't make it up the mountain alive. It was a very clear threat and, considering the source, Becker took it seriously. Switzerland was Rich's territory, and anything could happen on the icy mountain road traveling up to Davos. But Becker hesitated to tell the group because he "was afraid of getting them afraid."

This was not the only threat. In addition to the incident at the London Metals Exchange, Joe Uehlein had received phone calls at his home late at night threatening that his wife and young children would be hurt if he didn't back off the campaign. Someone had also tried to tamper with Becker's car in Pittsburgh. As Becker tells the story,

> The police here in Pittsburgh, they didn't apprehend a person, but someone was caught trying to do something in my vehicle, cutting the wires. They actually set the alarm off in the car. It was a stupid mistake they made. They had got in the car and were working in the car for a long time and apparently they were trying to connect something up. The police said they apparently weren't trying to steal it, possibly trying to connect something up in the ignition. And the unasked question to me was, "We can't imagine why they would be doing this, can you?"

For Becker, the threats were becoming too real and too frequent. He decided the time had come for him "to deal with it in his own way." He couldn't get to Rich but he could get to Rich's people in the United States.

> After that, I got hold of what I consider the top principal here in the United States and told him . . . "Look, there's been two occasions where I've been told that violence was going to occur. You say you don't want this on your conscience. Well, that's not good enough. Maybe I can't get to a Marc Rich, maybe I can't get to a Willy Strothotte, but I can sure as hell get to you, and I want you to know that if anything happens to any of our people, directly, indirectly, accident, on purpose, it makes no difference, your ass is going to pay for it. It's already been taken care of. You understand that? It's already been taken care of, here, with you. So you better figure out some way to control this."

Somehow it worked, or at least something worked. The threat about the Davos trip was the last the campaign received.

Meanwhile, the union delegation left for home, unaware of how close they had come to confronting Rich's power at its most ruthless. Although they were frustrated that they hadn't been able to pull off a face-to-face meeting with Rich or Strothotte, they returned with a fistful of newspaper clippings. They eagerly awaited the *Rolling Stone* article, "NBC Dateline" footage of Garment and Reynolds locked out of Rich's headquarters, and shots of Brian Ross surprising Rich on the ski slopes of St. Moritz.

NBC aired its story about Marc Rich on February 12, on the nightly news. To the union's dismay, the Ravenswood struggle was not mentioned. But the story was filled with bad publicity for Rich, describing him as one of the world's most notorious white-collar criminals and reporting that authorities were now raising questions about his business dealings with Saddam Hussein before, during, and after the Gulf War. Whether NBC mentioned it or not, Marc Rich surely knew that Brian Ross and his crew would never have tracked him to Switzerland, nor mentioned him on the evening news, if it hadn't been for the union campaign.

The next stop on the tour was Romania, one of Rich's strongest bases in Eastern Europe, where he had long ago cultivated close ties with dictator Nicolae Ceausescu. After Ceausescu's ouster, Rich's business arrangement continued with the new leaders, most of whom had been closely allied with Ceausescu. By 1992 Rich had not only expanded his petroleum and metal trading interests in Romania but, as the union had discovered the previous fall, he was in the process of buying 51 percent interest in the Athenée Palace, one of Bucharest's largest and oldest luxury hotels, from the Romanian government, who would retain the remaining 49 percent.

In January, Rich Yeselson had met in Washington with Miron Mitrea, president of FRATIA, the largest independent union federation in Romania, to talk about Ravenswood and learn as much as he could about Rich's investments in Romania. Yeselson found Mitrea to be a willing ally, for the Romanians deeply distrusted Rich, viewing him as an opportunistic parasite, eager to feed off his close ties with Ceausescu's former allies in the new government. Even though Mitrea and other union leaders felt a desperate need for more foreign investment, they were not interested in investment from the likes of Ceausescu's former friend and business partner, and they eagerly offered their support for the Steelworkers' campaign.

Despite these initial connections, Penny Schantz and USWA international affairs director Jerry Fernandez soon discovered that, unlike in Switzerland, where most of the union contacts spoke English and where Joe Lang eagerly took on many of the logistical details, arranging the trip to Romania was extremely difficult. The phone and fax connections were primitive. It took Schantz more than three-and-a-half weeks just to make her first phone connection to FRATIA and six weeks to work out arrangements for the trip.

The plan was for the union delegation to speak at a large labor rally on February 14, 1992, in Bucharest's main square, at the same time that similar rallies were scheduled in forty-five other cities throughout Romania. The rally, organized by a council of Romanian labor federations including FRATIA, was supposed to focus on high unemployment and the rapid disintegration of social protections for Romanian workers. When a week before the rally Penny was still having difficulty getting through to FRATIA, the strategy team decided that it was too expensive and difficult to make all the visa and travel arrangements to send a worker delegation. In the end they decided that Penny would give the speech and Tavia would join her with the puppets, while FRATIA staff would help them distribute thousands of copies of the Marc Rich "Wanted" poster, translated into Romanian.

Some in the Steelworkers organization began questioning the efficacy and cost of dragging the two puppets across Eastern Europe. Penny in particular wondered if they were more trouble than they were worth. Dewey Taylor, too, had found the puppets problematic at first. But, struck by Tavia's spirit and commitment, he had spent the two weeks between the trip to Switzerland and the trip to Romania trying to iron out some of the most unworkable aspects of the puppet's design.

The International, I think, started having some second thoughts but . . . I convinced them that I could do something. I saw what she had, and I took the tall pole that made Mother Jones and made three pieces out of it, made a slip joint that would go together. Went to Parkersburg and bought sheet aluminum. One of our fabricators out of the plant that has a weld shop out of town, took it to him and told him what I wanted. And he bent it and made two aluminum boxes, put wheels on the bottom of the boxes, coaster wheels that we could put inside, and when we got there we could take the wheels out of the box and put them back on. And so we had two much smaller boxes. The airlines told us what size the boxes really had to be to get them on with a passenger.

All went well until Tavia got to the Romanian border. Despite Dewey's handiwork, the custom agents balked at letting the giant Mother Jones puppet into the country. Anticipating problems, Tavia had stuffed the Marc Rich head in her backpack, putting her clothes inside the puppet. So "Marc Rich" made it through undetected.

The rally was a huge success. According to press reports, more than twenty thousand people gathered in the main square in Bucharest while another thirty thousand rallied in smaller cities across Romania. Penny gave a rousing speech while Tavia danced around her on stage wearing the Marc Rich head. By the time Penny was finished, the crowd was booing Marc Rich and cheering on the Steelworkers. According to FRATIA's English-language publication, Schantz had convinced the crowd that the Romanian people should unite "with Ravenswood workers and not allow [the] Romanian government to make business with this dangerous man."

The rally and Schantz's speech received extensive coverage in the Romanian media, with repeated footage of Marc Rich being booed on national television. Shortly after, through pressure from the Food, Agricultural, Hotel and Restaurant Secretariat (IUF), the Romanian government intervened to strip Rich of his interest in the Athenée Palace Hotel. As with the Slovakian smelter just three months before, this action proved to Rich that the union had the power to interfere with what he did best—lucrative deal-making in the newly emerging market economies of Eastern Europe.

Next the tour moved on to Czechoslovakia, where the Steelworkers hoped to build on their earlier success. Because of his close contacts with the Czech metalworkers and CSKOS, the Czech/Slovak labor federation, the strategy team decided that Jerry Fernandez would coordinate the trip. In fact, Jerry's connections with the metalworkers had led to the original tip about Rich's impending purchase of the Slovakian smelter. Instead of a rally, the metalworkers suggested a press conference and, instead of the puppets, which they felt were too theatrical, they asked that Jerry bring with him a locked-out worker.

On February 27, Jerry Fernandez, Penny Schantz, and Charlie McDowell traveled to Prague. For McDowell, Prague was a different and curious experience.

> Well, it was like watching a movie of Europe, Russia. We flew into Prague. They sent a limousine and a driver. They picked us up. They delivered us to a hotel, an old hotel, a state-owned hotel that was staffed by the workers that worked there before and during communism. And, hell, the room I had was bigger than this house I live in. Huge. Huge. Huge. An old hotel, but nice. And you went down to eat . . . they didn't care whether they waited on you or not, because they were still under the state of communism. But they were nice and the food was just European food, to tell you the truth. And if we went anywhere, the limousine came and picked us up, took us where we was going, and delivered us back. And there we had an interpreter, I believe it was the secretary of the Czech metalworkers.

At the press conference, Jerry consciously took a back seat and let Charlie do most of the speaking, "Charlie did a great job. The media loved him. He was colorful. He said things certainly in a manner in which a worker would say, yet very articulate at the same time. I don't think that, had I myself gone there and tried to spread the word, I would have gotten the same press coverage. There'd have been a press conference. I'd have met the same people, but the impact wouldn't have been nearly as great."

Just as he had in Switzerland and the Netherlands the summer before, Charlie moved and captivated the crowd. But, unlike when he spoke to the Swiss reporters or Dutch bankers, this time he appealed to his listeners as fellow workers and trade unionists: "So we go out there, and all these mikes here, and back there they had those interpreters behind the glass. Then, Jerry just said a couple of words to them, and I started describing what was happening, why I was there . . . because [Rich] was a fugitive from the United States, a wanted criminal, owning a plant in the United States that I worked at, trying to break a union. And I was emphasizing the fact that we were an organized union the same as the Czech Metalworkers were, and that I was there to make them clearly aware of what this individual was doing to the plant that I worked at and what would happen, I felt, if he bought [the smelter]."

McDowell and Fernandez were joined by the president of the Czech Metalworkers and a delegation of workers from the Slovakian aluminum smelter that Rich had tried to purchase. As in Romania, the event garnered extensive press coverage, including one newspaper headline calling on the Czech government to say "NO" to Marc Rich. The union delegation spent two more days in Czechoslovakia meeting with metalworkers, urging them to continue their fight to keep Rich from conducting business in their country.

While Jerry and Charlie flew home, Penny went to London, where she met up with Joe Uehlein, Dewey Taylor, and Tavia LaFollette for the next leg of the tour. Following up on contacts they had made during the trip to the London Metals Exchange in October 1991, the union planned to picket outside Rich's London headquarters and speak at a press conference hosted by the British Trade Union Congress (TUC). Because of the threats and harassment they had encountered on their last trip to London, the union delegation was somewhat nervous about returning to Rich's headquarters. But they were reassured by their friends at the TUC that the police would be watching and, as long as they remained on the sidewalks in front of Rich's headquarters and avoided a direct confrontation, they would be fine.

This time the puppets came together without a hitch and attracted a great deal of attention outside the headquarters. "The average person on the street, of course, didn't know what was going on. They had no idea what this represented. And they were only going to read about it in the newspaper the next day," recalls Dewey. "But that was the whole idea, the newspaper and the television people, to get the exposure. And I think we did a good job of getting television and newspaper exposure. But the people on the street would kind of go by [and] go, 'What in the world is going on here?'" While they were leafleting and, as Uehlein liked to say, "puppeting" outside Rich's headquarters, they were joined by representatives from a London public relations firm who distributed a "fact sheet" on behalf of Marc Rich, once again disclaiming any connection to the Ravenswood dispute. For the union delegation it was one more piece of evidence that they had put Rich on the defensive.

The press conference that afternoon was chaired by Norman Willis, the TUC General Secretary. The British equivalent of the AFL-CIO president, Willis brought a great deal of stature to the campaign. But, as in other campaign events, a locked-out worker, in this case Dewey Taylor, stole the show. As he says, by then he knew how to "tell the story well enough to be a little bit dramatic about it and kind of impress people so that they could almost sit there and sense what was going on. They could see a terrible injustice being done to a bunch of people, and all in the interest of making money." The British press coverage was fairly limited, but the story was carried at length in trade publications such as *American Metal Market*, read by Rich's colleagues and competitors in the industry. The Steelworkers were also pleased with the support they received from the British labor movement.

The next day Uehlein was joined by Michael Boggs for a special meeting with Matti Huhanantti, a Finnish oil trader, yet another of the many enemies and competitors of Marc Rich. Hougan had come across Huhanantti in his research into Rich's infamous dealings with the Finnish national oil company (NESTE), which had exploded onto the front pages of the Finnish press in fall 1991. At Huhanantti's request, Uehlein and Boggs met with him at his London home, where, according to Uehlein, "Huhanantti spun out an amazing laundry list of charges against Rich and made an even more amazing offer—to broker a contract with British Special Forces operatives to apprehend Rich, but at the cost of over a quarter of a million dollars, and the possible loss of human life."

Uehlein and Boggs quickly rejected the offer and ended the meeting. As they left Huhanantti warned them that if Rich were to learn of their meeting he would have Huhanantti murdered. Just one week after their meeting they heard that Huhanantti, a world-class skier, had been hit from behind on the ski slopes and had broken his neck. Although they never were able to confirm what happened, they worried that Rich's forces might have engineered the "accident."

While Uehlein and Boggs were meeting with Huhanantti, Penny, Dewey, Tavia, and the puppets flew to Amsterdam, where they were joined by Rich Yeselson. Through Penny Schantz's contacts, the Dutch trade union federation, the FNV, offered their support in setting up a press conference and a rally outside Rich's Rotterdam headquarters.

After scouring Amsterdam for a van big enough to hold the puppets, they rented a small station wagon that barely fit the four of them and the new metal boxes Dewey had crafted for the puppets. Dewey ended up driving. "We couldn't find an automatic, and none of the others said that ever in their life had they ever driven a stick shift, and they said if we take this vehicle, the driving's left to you. And we took a vehicle and went out trying to find our way, and of course . . . none of us could read the signs. None of us knew the language. And so we had quite an experience with me trying to drive and three backseat drivers helping me." Somehow they made it to Rotterdam, where they leafleted outside Rich's headquarters and held a press conference sponsored by the Dutch union federation, FNV. Unlike in London, they garnered a great deal of press attention in both Amsterdam and Rotterdam, with large photographs of the puppets, outlining the story of the Ravenswood lockout and Rich's connection through Rinoman and the NMB Postbank.

After Rotterdam, Schantz and Yeselson went to Paris for the last stop on the European portion of the seven-week, seven-country "Marc Rich Tour." The French metalworkers' leaders, affiliated with the Force Ouvriere, were clearly uncomfortable with the puppets, arguing that the French would simply laugh at "this giant marionette." So they left the puppets behind and simply leafleted outside Rich's headquarters on the Champs Élysées. But the French newspapers gave the campaign thorough coverage, including an article in *Le Figaro*.

The last international action started March 15 in Caracas, Venezuela, at the Congress of the International Confederation of Free Trade Unions (ICFTU), a gathering of union leaders from all over the world. The previous summer, the Steelworkers had been able to eliminate Rich from bidding for a billion-dollar smelter deal with the Venezuelan government through the assistance of the metalworkers and the Venezuelan national union federation. Now the union hoped to build on that victory and use the ICFTU Congress to spread the message to as many unions in as many countries as possible.

Penny Schantz and Jerry Fernandez worked with the ICFTU to have Rich "Wanted" posters printed in as many of the languages represented at the Congress as they could. The Steelworkers hosted a press conference featuring Lynn Williams, AFL-CIO secretary-treasurer Tom Donohue, Joe Uehlein, and ICFTU and Venezuelan labor officials. Meanwhile, Mother Jones and Marc Rich drew large crowds, as Tavia and the puppets made their last stop on the tour before she had to return to her studies.

Tracking Rich across seven countries in seven weeks, the union had, in USWA organizing director Bernie Hostein's words, "put up a picket line around the world." As Joe Uehlein told *The Nation* reporter David Corn, "Marc Rich is not a quiet commodity anymore." The Marc Rich Tour had ensured that the glaring spotlight of publicity, which Rich sought at all costs to avoid, was shining on him as never before.

19

BOYLE'S RETREAT

Flush with their victories in Europe and at home, on April 15, 1992, the strategy team gathered for a meeting with George Becker at the Steelworkers headquarters in Pittsburgh. They had a lot to talk about. Hougan and Yeselson had just finished the first draft of a new campaign newsletter, "The Rich Report," for regular distribution to the media and their labor allies around the world. With the support of the Seafarers and their fellow unions in the International Transportation Federation, the campaign was gearing up for sympathy picket actions against ships handling Rich's business at ports around the globe.

There was big news from Bulgaria, where more than ninety thousand coal, uranium, lead, and zinc mine workers had just returned from a one-week strike focused, in part, on a secret deal between Marc Rich and the Bulgarian government in which Rich had purchased millions of dollars' worth of lead at a fraction of its cost. Schantz, Fernandez, and Uehlein wanted to quickly finalize details for a trip to Bulgaria to capitalize on the anti-Rich sentiment generated by the strike. International actions were also being planned for May and June in Russia, Australia, Finland, Spain, Hong Kong, and Israel.

In the United States, the Wise Committee investigation was in full swing in Congress, while the West Virginia legislature was ready to start its own investigation into the Ravenswood dispute. The environmental campaign was rapidly picking up steam, and pressure was mounting on Pepsi and Lone Star Beer to stop their purchase of RAC aluminum. Judge Ries was expected to issue the NLRB decision any day, and the Steelworkers' appeal of the gag order surrounding Boyle's divorce case was due to be heard June 2 in the West Virginia Supreme Court.

The group was also excited about Scott Spencer's cover story in *Rolling Stone* magazine, which had hit newsstands the day before. In between articles on Def Leppard and Bruce Springsteen appeared an eight-page spread of photos and text titled "Hope and Hard Times." From the picket shacks in Ravenswood to Tavia and the puppets in Zug, Spencer told how "for the men and women forced from their jobs at Ravenswood

Aluminum, the struggle to regain their future took them from small-town Appalachia to the glass canyons of international finance." Fifty-year-old locked-out workers, their wives, and their friends and neighbors, most of whom had never before even considered purchasing *Rolling Stone*, now lined up at Foodland, Rite Aid, and 7-Eleven for the few copies available. Some drove to Parkersburg or Charleston, but most came home empty-handed. There just weren't enough copies of *Rolling Stone* to go around.

But when Becker joined his team at the large round table in the conference room adjacent to his office, he announced that he had bigger news to discuss than *Rolling Stone* or the trip to Bulgaria. Willy Strothotte had seized majority control of Ravenswood Aluminum Corporation and in a few short days would force Emmett Boyle out. Boyle would be replaced by a new CEO and company president hand-selected by Strothotte and Rich. Furthermore, Strothotte had hired attorney Peter Nash, former general counsel of the NLRB, with instructions to restart negotiations with the Steelworkers and reach an agreement as soon as possible. Pending the outcome of these negotiations, the entire corporate campaign against Rich and Ravenswood, both in the United States and abroad, would continue, but the escalation would stop.

The team members were stunned. It was incredible, wonderful news. They had not only beaten Emmett Boyle, they had forced Strothotte and, behind him, Marc Rich, to the bargaining table. At the same time, the news was also quite confusing. Here they were straining at the bit to jump into the next phase of the campaign, to, in George's words "escalate, escalate, escalate," and suddenly they were being told that, as a result of months of secret, behind-the-scenes exploratory discussions, their role in the strategic campaign they had been living and breathing, day and night, for more than a year, might be over. How were they going to put the brakes on a campaign that had taken on a life of its own?

Becker explained that the negotiations grew out of informal conversations he had held off and on with Leonard Garment since that first meeting in Garment's office in spring 1991. Garment had offered himself as a middleman, primarily to carry signals and messages back and forth from one side to the other. Then, in late December 1991, Becker heard from his longtime friend, New York attorney Gene Keilen. Because Keilen had connections in the labor movement, he had been approached by another bankruptcy attorney, Leon Marcus, who did legal work for Clarendon as a lawyer in Milgrim, Thomajon, and Lee's bankruptcy division. According to Becker, Marcus and several associates wanted to join the bankruptcy group representing the *Daily News*.

They wanted to be the lead lawyers in the *New York Daily News.* And when the union there found out . . . that his group now had associated themselves with Clarendon, they said, "That's Marc Rich and we aren't going to deal with you." And Leon Marcus just went crazy over this because he saw millions of dollars in fees slipping away. And he also saw that he could wind up blacklisted by labor. Since a lot of the deals he worked on were through folks like Gene Keilen, who represented labor, he picked up the phone and called Gene Keilen and said, "I think I could help settle Ravenswood." He says, "I'm speaking without authority," but he really wanted to broker this. . . . "What do the Steelworkers really want? What do they really need? I have connections back into the Marc Rich headquarters."

Keilen checked with Becker and, although they were skeptical at first, they quickly discovered that Leon Marcus did indeed "have connections back into Marc Rich headquarters." Marcus worked closely with a lawyer who, in turn, was a close associate of Robert Thomajon. Thomajon was the attorney who had originally negotiated the Ravenswood buyout deal in 1989 and subsequently quit his law practice to work directly for Marc Rich and Willy Strothotte. So, early in January, Becker and Keilen spent an afternoon with Marcus in Keilen's office, talking over what it would take to settle the lockout. For Becker the main issues were getting rid of the scabs, dropping the RICO suit, and getting every locked-out worker back into the plant. Marcus relayed that Strothotte and Rich wanted guarantees that the public hearings would end if a settlement was reached, and that there had to be some way for the company to save face in the final settlement. According to Becker, "I told him that's crazy. He'd better worry about the union saving face. If the union doesn't win this struggle, we'll be so close to Marc Rich he'll think he's married to us."

When Marcus added that they couldn't take all the locked-out workers back because they needed to "leave room for some of the good scabs the company wanted to keep," Becker was struck that this was very similar to the language Garment had used in a conversation a few weeks before. This convinced Becker, that Marcus, like Garment, was working under Strothotte and Rich's direction and authority.

Over the next few weeks there were more conversations back and forth between Marcus and Keilen, but nothing materialized. In early February 1992, not long before the AFL-CIO's annual executive council meetings in Bal Harbour, Florida, and just after the union delegation returned from Zug, Becker received a call from AFL-CIO president Lane Kirkland, telling him that he had just received a call from Leonard Garment. Kirkland had a relationship with Garment, going back to the 1970s when they both worked on Patrick Moynihan's Senate campaign. Kirkland reported that Garment wanted him to set up a meeting, while they were down in Florida, with Becker, Kirkland, Garment, Lynn Williams, and Willy Strothotte.

So, in the same week that twenty thousand Romanian workers booed Marc Rich at the FRATIA rally in Bucharest, and less than two weeks after Rich's security guards had handed Dewey Taylor and Mike Bailes Strothotte's note declaring that he was not going to get involved in the Ravenswood dispute, Willy Strothotte traveled from Zug to Bal Harbour, Florida, to meet with the Steelworkers.

Over six feet tall, blond, and cosmopolitan, Strothotte appeared to Becker "a very impressive guy . . . the kind of person . . . if he walked into a crowded room, he would be one of the people that everybody's attention would sort of just gravitate to." But despite Strothotte's confidence and charm, Becker found their first meeting confusing and disappointing. Rather than a discussion like they had had with Marcus about firing the scabs and negotiating a settlement, all Strothotte wanted to talk about was how to sell the plant to the Steelworkers under an employee stock ownership plan (ESOP). Confused about the mixed messages, Becker called his friend Gene Keilen, who just happened to be vacationing in Florida with his family. Keilen also happened to be one of the leading advisors to the labor movement on ESOPs. That night Keilen and Becker set up another private meeting with Strothotte in West Palm Beach, where Strothotte was staying.

> We had a pleasant dinner and we talked hardball. And Gene wanted the company's financial records and all this stuff to look at it. In time, Strothotte said, "Bullshit, there aren't any financial records, what are you talking about, financial records?" And Gene said, "There's financial records on everything." And Strothotte said, "Look, you've got a pension fund. Just dip down in the pension fund, tell us what you'll give us for it." And Gene says, "Well, how can I tell you what we'll give you for it if we don't have any idea of the equipment that's saved and all the stuff there?" And [Strothotte] said, "Well, nothing's changed since we bought it."

Toward the end of the discussion, Strothotte accused Becker of acting as if the Steelworkers weren't interested in purchasing the plant. Becker says he responded, "Hey, wait just a goddamn minute. We *don't* want to buy the son of a bitch. We want to settle the contract, but if the only thing that we can do, the only game in town, is to buy it, we'll buy it rather than let it go under. . . . We'll put together an employee ESOP . . . but we've got rules and regulations on how we do that. . . . We don't have a war chest that we can just dig down [into] and buy a goddamn plant."

Late that night, as Keilen and Becker drove back to Bal Harbour from their meeting with Strothotte, they puzzled over the mixed signals they were getting. On one hand, Leon Marcus was talking about getting the scabs out and settling an agreement and, on the other, Strothotte, out of the blue, was pushing an employee buyout. Both Becker and Keilen hoped that Marcus would come through for them, but now, after listening to Strothotte, they were not sure Marcus had any authority to speak for Strothotte or Rich. They felt they had to keep a dialogue going with Strothotte, because they didn't want the plant to shut down, and if an ESOP was the only way to save the plant, they would have to keep that option open.

At this point, though, Becker was convinced that they needed to somehow redouble the pressure on Rich and push the dialogue toward getting the scabs out and the union workers back in. "My message to [Rich] and to Strothotte that night was, 'Look, you think you've had hell from the Steelworkers union so far . . . you ain't seen nothing yet. If you shut that plant down, all you've done is to eliminate any possibility of settling with the Steelworkers . . . We'll be like a cancer eating away at you forever. . . . We'll not give up on this thing.' You better settle the son of a bitch some way. . . . Don't you shut the plant down, [because if you do] I don't see any way for us to ever settle this dispute or for us ever to stop the fight.'"

Over the next several weeks Becker escalated the campaign, never letting on to anyone else on the union strategy team that he had been in touch with Strothotte or that he felt the pressure campaign was working. He was not only afraid that news of the purported breakthrough would undermine the zeal for the campaign, but he also feared that, if a leak got back to Boyle or Rich, the entire effort would be sabotaged. Meanwhile Gene Keilen kept in touch with both Marcus and Strothotte, still trying to push things toward a settlement rather than an ESOP.

Then in early March, just as the union wound up the seven-country "Marc Rich Tour" and the Wise committee announced its intention to step up efforts to arrest Rich and bring him back for trial, George Becker received another call from Lane Kirkland—Garment wanted a meeting. Kirkland told Becker that he sensed the union was making some "real headway on Ravenswood."

So another meeting was set up with Strothotte, this time in Garment's office in Washington, D.C. Once again Garment asked that Kirkland and his assistant, Jim Baker, come along with Becker and Williams, both because Garment was comfortable with Kirkland and because Rich and Strothotte appeared to believe that Kirkland brought more status to the negotiations. Unlike at the previous meeting, Strothotte and Garment brought with them an entire team, including Rich's attorney and former number two man in the Justice Department, Brad Reynolds; Craig Davis, former executive vice-president of Alumax Inc.; and Pete Nash, who had been NLRB general counsel during the Nixon and Ford administrations. Becker knew that Davis had been working for Rich out of the Zug headquarters since he had left Alumax two years before and that he had played a key role in negotiating Rich's interest in the Alumax smelter in Mount Holly, South Carolina.

Nash's connection to Rich and Ravenswood, if any, was less clear. Since leaving the Board he had represented employers in some of the nation's biggest NLRB employment discrimination cases. His most recent case had involved ConAgra, a large Midwestern meatpacking conglomerate, where he had brokered a $40 million NLRB "discrimination for union activity" case down to $7.1 million. Whatever Nash's connection, Garment and Strothotte made clear from the beginning that Nash was there at Marc Rich's behest, to do what was necessary to get a settlement.

At first Davis and Nash kept quiet, leaving most of the talking to Willy Strothotte. According to Becker, Strothotte took on a different tone than he had during the first meeting in Florida. He started out by stating that they wanted a settlement and looked to Becker, asking, "What do you want to settle?" Everyone in the room turned to Becker for a response. "And I said, 'The goddamn scabs all have got to be fired, every one of the scabs have to go. Every one of our people have got to come back to work, those that have been fired, those that are on the line. We've got to be compensated, back pay, for everything that we've lost.'"

Although in his previous discussions with Garment, Marcus, and Strothotte, Becker had been extremely firm that every scab had to be removed, this was the first time he had focused on the back pay. He knew that the Board decision was in the wings, and Rich Brean and Stuart Israel had convinced him the union had a good shot at winning, so he decided to call their bluff. At first his remarks were met with stony silence. Then, when Garment questioned what business he had raising the back pay issue when it had never been mentioned in earlier discussions, Becker responded, "Hey, you had your chance. . . . You fucked it up. If you had settled it then, we would have had a chance to reach an agreement. You've waited too long. We're now facing the Board decision. . . . We've got to deal with that issue. . . . And, if you're not interested, then, as far as I'm concerned, we'll continue the struggle."

Garment was clearly furious at Becker's response and for a moment Becker wondered if the discussions would break off then and there. But then Pete Nash jumped in: "We'll deal with that. No promises, but we'll deal with that." Then, almost as an afterthought, Becker threw in that Emmett Boyle had to be fired. "There was too much bitterness and too much background on his personal involvement, lying, deceit, misleading the people, and a complete fixation on his part that the union had to be destroyed, and there was no way that the people could accept and work with him in any credible way. He had to go." Becker was again met with silence, and was concerned that Strothotte and his team would simply stand up and walk out of the room. But then, according to Becker, "Pete Nash spoke up and said, 'You're right. . . . With as much turmoil and as much problems as they've had down there, he's got to be a major factor in the whole thing.' And he turned to Craig Davis, and he said, 'And this will help clean the slate . . . so that a new management can start off fresh.' So we just disposed of Emmett Boyle right then, with about two dozen words."

Suddenly the meeting took on a whole new dimension. Becker had walked into Garment's office that afternoon confident that the campaign was taking a toll on Rich, but still with very little idea or expectation of what, if anything, could be accomplished in the meeting. And now, moments into the discussion, they were, in Becker's words, "almost settling the future of Ravenswood right there, [in] just a few casual minutes."

There was still much to be done before the union and the company could once again sit down at the bargaining table and negotiate an agreement. Strothotte and his group had to get majority control of Ravenswood stock, they had to remove Boyle, and they had to reconstitute the company. But, according to Becker, the difference was that now they "had a framework to sit down and work towards."

After the meeting, while the rest of the group were getting acquainted, Nash and Becker went off to the side to speak privately. They talked about Boyle and they talked about how they were going to make this happen. As Becker remembers, "We talked about how we'd work together and what I could expect from him and what he could expect from me and we just sort of pledged ourselves, that if it worked, OK, if it blew up, OK, but at least we would keep ourselves shooting straight on this thing."

In the weeks that followed, Becker got frequent reports from Nash on the progress his team was making in gaining controlling interest in RAC and removing Boyle. All the while, to hold Nash's feet to the fire, Becker kept pushing his strategy team to escalate the pressure against Rich and Strothotte, never once letting on about the progress being made toward a settlement. Several times Nash and Garment asked him to back off the campaign or turn down the heat on Rich, but Becker refused.

Yes, we kept escalating it, because it appeared to me that time was starting to drift by since the Strothotte meeting. They weren't coming to a conclusion on some of the really important issues. And they assured me it wasn't because they weren't trying to, and they wanted me to scale back our campaign as a show of good faith. But I took the position that, "Look, we can't scale back . . . One, I would have to explain to our people why we're doing this, which we don't want to do at this point in time. Second . . . if we stop anything and this falls apart, it's going to be much more difficult to start back up, and I'm just not going to do that. You're going to have to accept the fact that we're in a war situation, and the faster you can move your principal toward a settlement, the faster we can cease hostilities."

Becker pushed Jim Hougan to gather further information on Rich and his business transactions from Rich's former clients and associates. Through Hougan's interviews with federal marshals, FBI agents, disgruntled clients, former business partners, and even convicted felons, the dossier on Rich's wheelings and dealings grew thicker and thicker. Not wanting to jeopardize the emerging negotiations but still wanting to keep the pressure on, Becker made the decision to selectively, but anonymously, leak some of the information they had collected directly to Rich—information that would most likely be very damaging, but that had not yet been used in the campaign. This way they were able to signal to Rich both the union's current restraint and its potential for future action should progress toward a settlement break down.

By the end of March, with Judge Ries's decision due out within the month, Becker grew concerned about Nash's lack of progress in engineering Boyle's ouster. He knew that once the NLRB decision, with its potential for millions of dollars of back pay, became public knowledge, the workers would refuse to settle and RAC would most likely be facing bankruptcy. Nash tried to reassure Becker that they were close, but said they had to hold off making their move because Boyle was so irrationally committed to breaking the union and retaining the scabs, that he might, in Nash's words, "kick the traces" on the way out. Some of the scabs had brought guns into the plant and might do a lot of damage if management settled with the union. By shooting out the power system, in just a few hours they could freeze up the potlines, causing millions of dollars' worth of damage. Before Boyle got wind of his impending ouster, Nash needed to line up a replacement management team and to set up a backup security system to control the scabs.

Once majority control of the RAC board was achieved, the plan was to have Strothotte call Emmett Boyle to Zug for a meeting. Boyle would be told pointblank that, as far as Strothotte was concerned, the Ravenswood dispute needed to be settled right away and there were simply too many hard feelings for that to happen with the present management. Strothotte would offer Boyle a face-saving way out, but if he refused, they would publicly force him off the board and the management team. They assumed, however, that Boyle wouldn't go without a fight and might very well try to bring the plant down with him.

Already Boyle had grown suspicious that Nash and Strothotte had been in contact with the union; once he directly confronted Nash on whether he had spoken to Becker about the RAC contract. Nash vehemently denied that he had ever talked to anyone in the union about the contract. Later he reflected to Becker that that was true, given that all of their discussions to date had focused on how to remove Boyle and end the lockout, not on actual contract negotiations. Still, time was running out.

On April 8, 1992, Nash and Strothotte finally made their first move. Willy Strothotte notified Charles Bradley that he was going to exercise his legal right on behalf of Rinoman Investments to take control of Bradley's 20 percent share in Ravenswood Aluminum. Strothotte was able to take this action because Bradley had defaulted on an $11.5 million loan from Rinoman in 1991, for which he had used his 20 percent share in RAC as collateral. Adding Bradley's 20 percent to the 48 percent share Strothotte already had in RAC through Clarendon, Strothotte could use Rinoman's 68 percent control to vote to expand the board to include two new officers, Craig Davis and Jean Loyer. Davis coordinated worldwide aluminum operations for Marc Rich. Loyer was the former president and CEO of Howmet Corporation, an aluminum producer bought out by Alumax in the early 1980s. With Strothotte, Davis, Loyer, and Clarendon senior financial officer In-Suk Oh now representing Rinoman Investment's shares in RAC, Strothotte had the majority he needed to force Boyle off the board.

The new, expanded board held its first meeting on April 13. Although notified of the meeting, neither Boyle nor Bradley attended. At the meeting the four directors took the following actions: they removed Boyle as RAC chairman and CEO and replaced him with Craig Davis; they removed Bradley as secretary; and they appointed a Special Labor Committee composed of Davis and Loyer to "exercise all powers of the full board to attempt to resolve the current labor dispute."

That same day both Bradley and Boyle sent angry letters to Strothotte and In-Suk Oh, contesting their right to vote Bradley's common stock and expand the RAC board. Boyle's argument had an interesting twist. Calling Rich a "notorious fugitive from justice," the letter concluded, "Considering the damage the Corporation has already suffered by virtue of having been wrongly accused of being controlled by Marc Rich, imagine the harm that will occur once this accusation becomes fact."

The next day, to defend against Boyle's and Bradley's opposition, the newly expanded RAC board filed suit in the Delaware State Court of Chancery to get court affirmation of the Rinoman takeover, claiming that because of Boyle's intransigence in resolving the labor dispute, RAC was now "on the brink of financial ruin." The suit charged, "Once a healthy and profitable company, RAC now is in default on its bank credit facility and will be unable to pay this $71 million obligation when it becomes due on June 5, 1992. . . . RAC's expanded board believes that efforts to seek a prompt and rational resolution of the labor dispute will materially assist RAC in obtaining an extension of the June 5, 1992, repayment date and thereby deferring an obligation RAC cannot pay."

The suit claimed that prior to the labor dispute the seventeen hundred workers at Ravenswood produced approximately 50 million pounds of semi-fabricated aluminum products per month, for annual revenues of $701 million. Since the lockout, the one thousand replacement workers produced only about 35 million pounds per month, and annual revenues had dropped to $491 million. The suit also mentioned the potential back pay award and said the company was "in default to its principal creditors." Finally the suit asked for a restraining order to keep Boyle and Bradley from "obstructing or interfering in the actions of the RAC board or from taking any action that would be inconsistent with any of the actions taken by the board."

Although the Rinoman suit was sealed, Nash and Becker knew it wouldn't be long before Boyle started arguing his case in the press. Together they prepared press releases and then Becker called his strategy team together. After telling the story and answering their many questions, Becker laid out the following ground rules to his team: everything was to be kept confidential until the press release two days later on April 17; everyone should be cautioned against deviating from the text of the press release and future press releases; all Marc Rich activities were to be kept on "hot idle"; no new initiatives were to be started; groundwork should be laid for eventual withdrawal from more visible projects such as the Senate investigative committees and the environmental lawsuit; and all the hotline information from that day forward needed to be cleared first with Becker and then the legal department.

It was a lot to swallow all at once. For Joe Uehlein, it felt like the campaign strategy that they had put together so painstakingly over the last year, like pieces of an extremely large and unwieldy puzzle, had finally come together. After the gambles and the doubts, the more controversial and innovative elements of their strategy, such as the "Marc Rich Tour" and the environmental campaign, had been vindicated. Uehlein was certain that it was the combination of the international campaign against Rich, coupled with the economic and legal pressures against Boyle and RAC in this country, that made it possible to talk settlement that April morning.

> Because by April, you've got to think about where Marc Rich is. What's going through his head at this time. He's lost big deals in Czechoslovakia and Venezuela. Everywhere he goes he's seeing Ravenswood. . . . A resolution in the European parliament blocking Switzerland's ascension to the E.C. because of Marc Rich. . . . He lost the Atheneé Palace Hotel . . . because of our communication. So this guy's sitting there thinking this is becoming real. For a thirty billion-dollar-a-year guy, this is a problem. . . . He's thinking this guy Emmett Boyle has sold us a bill of goods on this union-busting strategy. It's not working out. He's losing his customer base. And now he didn't even file his EPA permits. So these guys, both Leonard Garment, Marc Rich, anybody involved at any level of intelligence, [are] going to look at Emmett Boyle as a buffoon. . . . Didn't file his permits—let's get him out of there.

And now, somehow, they were going to have to put it all on hold. For the next two days, it took everything they had to keep the news to themselves until Becker sent out the press release.

Late in the evening on April 16, Judge Ries announced from his home in Washington, D.C.: "I've completed my decision. It should be issued within ten days." He refused further comment on the case until the Steelworkers, RAC, and federal officials had a chance to review the decision.

By early the next morning, the media had received a fax from Steel-worker headquarters: "We understand that R. Emmett Boyle, Chairman and Chief Executive Officer of Ravenswood Aluminum Corporation (RAC), is in the process of resigning. We also understand that the Board of Directors for Ravenswood have engaged Pete Nash, former General Counsel for the National Labor Relations Board, to explore settlement possibilities of the current labor dispute with our union. Needless to say, the United Steelworkers of America welcome both of these actions. We stand ready to commence negotiations immediately and to work toward a mutually satisfactory agreement and the prompt return to employment of our locked-out members."

Celebrations swept through Ravenswood and surrounding communities throughout the day as local newspapers broke the stories of Ries's impending decision and Boyle's ouster. Beyond a terse acknowledgment that "there are developments taking place," company officials had no response. Reporters who called the plant were told that all the RAC managers, including Boyle and Boger, were unavailable due to the Good Friday holiday. Following Becker's directions, Local 5668 leaders referred all press calls to the international headquarters.

Four days later, Becker told the press that he and Pete Nash had met in Pittsburgh that morning to lay the groundwork for future negotiations. That afternoon Boyle for the first time issued a statement of his own: "As a result of actions taken by the new Board of Directors of Ravenswood Aluminum Company, I have been removed as chairman and CEO of the company. Although I have opposed these actions, I do not believe a shareholder battle is in the best interest of Ravenswood. Therefore, I have no alternative. I have to accept the decision of the new Board of Directors. Since I do not subscribe to the philosophy of this Board, I have resigned as a director and will no longer be a shareholder."

Boyle wrote to the salaried workers and the replacement workers, thanking them for their hard work and dedication. Jim Bowen sent Boyle a fax saying simply, "How does it feel to be permanently replaced?"

There was great joy at Fort Unity that night as locked-out workers and their families gathered for their weekly Tuesday dinner. Jim Bowen told the cheering crowd, "We've been here one day longer than Emmett Boyle. He's gone, G-O-N-E, gone." But he also cautioned the crowd that they still had a long way to go. "It's step one of a lot of steps. We just made one hell of a step for mankind right here—mankind and the laboring people. . . . This creation of a monster will not be resolved overnight. . . . We've been dealt a new hand, by a new dealer! Let's make the best of it."

Even Joe Chapman put aside his usual cautious optimism to declare, "This is an historic event in the annals of the labor movement. This is not the war but it is a battle won. I think we're on the right path, and will soon be able to begin repairing the damage done."

Outside the union hall the signs attacking Boger, Boyle, and the scabs were gone. When asked why the union had taken them down, Bill Doyle replied, "We see no reason to antagonize them now that we've won." But Bowen told reporters that night that the union was holding firm on its position that every last replacement worker would have to be removed from the plant: "We still stand the same on the replacement workers. We will not work with them in the plant. We will start from there with our negotiations."

Inside the plant, the replacement workers felt shocked and betrayed by the news of Boyle's ouster. He had promised them that their jobs were permanent and that the union strikers had been fired, never to return. Now, suddenly, Boyle was gone and their world had turned upside down. Over the weekend, representatives from the new management team had met with front-line managers. After outlining how the labor dispute had brought RAC to "the brink of financial ruin," they raised the question on everyone's minds:

> Which brings us to a real tough issue—can we settle without bringing back the striking union members and laying off our present workers? It is unlikely. Our present workforce, and you front-line managers in particular, have accomplished extraordinary things since November 1990. And you have all worked under very difficult conditions. If at all possible, we want to keep as much of our present workforce as possible. But we also have to be realistic. If we can't settle our labor dispute, RAC may soon be out of business—and nobody, including our present workers, will have a job.

Managers were instructed to encourage as many as possible of the replacement workers to stick with the company until the contract was settled. They handed out information promising that in negotiations they would try to "save at least some current employees' positions," and that, if they had to lay anyone off, they would provide a lump sum severance benefit of one month's pay, accumulated vacation, and three months' continued health insurance, as well as an opportunity "to be placed on a preferential hiring list . . . to fill job vacancies in the future."

Three days later, in a further management shakeup, president Don Worlledge announced his retirement. It was another blow for the replacement workers, for throughout the lockout Worlledge had been one of their staunchest defenders, repeatedly telling the media how much more productive and efficient they were than the union workers. Stidham told reporters the union was glad to see him go. "I think he's one of the most outspoken persons down there that opposed us. He supported the scabs and Emmett Boyle."

Bargaining was due to start again on April 29, 1992. Both Becker and Nash realized, however, that bargaining would never get off the ground if Ries released his NLRB decision as planned on April 26. Becker felt it would be extremely inappropriate for the union leadership to ask Ries to delay a ruling that his members had been anticipating for over a year and a half. So on April 23, Nash filed a motion on behalf of RAC asking Ries to delay issuing his decision until June 1, on the grounds that "issuance of the ALJ's decision during the negotiating process—whichever way it goes—will, in all likelihood hinder rather than advance potential resolution of the dispute." Ries agreed to delay the release for six weeks.

Bargaining was due to reopen in Pittsburgh. While Becker and the other union negotiators prepared to return to the table, Joe Uehlein, Jim Hougan, Rich Yeselson, Penny Schantz, and Jerry Fernandez did everything they could to put a hold on the international campaign. It was not easy. Calls had quickly been made to cancel Jerry Fernandez's upcoming trip to Russia and the May 13 rally in Bulgaria. A letter went out from Uehlein and Becker to campaign allies in other countries, from Switzerland to Romania to Venezuela, updating them on the latest developments. In most cases they were able to stop actions against Rich, or at least remove themselves as event sponsors or participants. In others, such as Joe Lang's long-planned town meeting about Rich in Zug, the meeting went ahead as planned, with Lang making sure to report the progress that was being made. But, as George Becker would say, everything remained on "hot idle," ready to take off at a moment's notice.

On April 28, the night before the first bargaining session was set to begin, "NBC Dateline" finally aired the long-awaited follow-up to the story about Marc Rich. Originally the union had been promised that on March 31 "Dateline" would run an in-depth investigative piece on Marc Rich and the Ravenswood campaign, with footage of Garment and Reynolds "locked out" of Rich's headquarters in Zug as well as of the locked-out workers at home in Ravenswood. A few days before it was due to be shown, NBC staff notified the union the segment was being indefinitely postponed. When the piece finally aired a month later, it was a very different story than the one the union had been promised. Entitled "Rich Man, Poor Man," the piece focused primarily on the contrast between the Marc Rich reporters had caught on the ski slopes of the Swiss Alps and the hardships faced by individual locked-out workers. All references to the union's strategic campaign had been deleted.

But as they prepared to go into bargaining the next day, the "NBC Dateline" show reminded the locked-out workers just how far they had come. They had taken on one of the richest and most powerful men in the world, and he, not they, had blinked first.

20

SETTLEMENT

On April 29, 1992, nearly eighteen months to the day from the moment that the gates first shut on the union workers at Ravenswood Aluminum, negotiations began again in Pittsburgh. This time the whole world was watching—Bob Wise and the U.S. Congress, the international and national news media, and financial investors worldwide. This was no longer a simple negotiation between a local union and a West Virginia employer.

The thousand scabs stood on the sidelines, blind-sided by the sudden changes in RAC management. Many still held on to the promise Emmett Boyle had made to them so often during the previous eighteen months: that their jobs were permanent and the union was gone for good. Others expressed bitterness that the union workers, after a year of attacking Rich all over the globe, were now negotiating with his surrogates. Becker brushed off suggestions that the union had made some kind of hypocritical turnaround on Rich. "Everything we've done in the last eighteen months was designed to bring us to this point," Becker told reporters. "Why they didn't do this eighteen months ago—twelve months ago— five months ago, I don't know. But I don't look a gift horse in the mouth, they are here today and we're ready to go."

The opening negotiation session was held on the union's own turf, at international headquarters in Pittsburgh. Although the union team consisted of only George Becker, Joe Chapman, Jim Bowen, and the local negotiating committee, nearly everyone in the building had been touched in some way by Ravenswood. They knew just how critical these negotiations were for the future of their union. This time, however, Becker, not Chapman or Bowen, would lead the union team. Given Becker's strong role in the discussions leading up to the negotiations, this made perfect sense. He was the person in control of the strategic campaign that had brought them back to the table in the first place and was the only one who knew of the understandings reached in the early delicate conversations with Garment, Strothotte, and Nash. Chapman, who had led the earlier negotiations, welcomed Becker taking the reins.

The local negotiators were very different men from the group who had sat down at the table in Washington, Pennsylvania, eighteen months before. They had taken on one of the most ruthless and wealthy financiers in the world and forced him, against his will, to the bargaining table. They had traveled to places they had never heard of to tell their story, lobbied Congress, testified in court, and made lifetime friends of union workers across the United States and in Europe. And they had comforted their frightened and grieving members and kept hope alive when many others told them to give up. All the while they had stuck together despite dramatic differences in personality, style, and ambition. The experience had both sobered and strengthened them.

But despite everything they had learned, the task ahead of them might be the hardest yet. Forcing the company back to the bargaining table was one thing. Reaching a settlement would be much more difficult. Unlike when they had started negotiations two years earlier, they were now facing a company on the edge of financial ruin, in large part because of the Steelworkers' campaign. They also had to accept that, in both the early court-ordered negotiations during the first year of the lockout and in off-the-record bargaining sessions during the last several months between Joe Chapman, Earl Schick, and the federal mediator, the union had already come down on its wage demands and signaled its willingness to give up the metal price bonus and make concessions on job combinations and other work rules. The major issues for the union were still on the table—pensions and summer relief and overtime in the pot rooms. But, unlike in 1990, the union would now also have to bargain to get every locked-out worker back in the plant, including those fired for felony offenses, get the scabs out, get the union back in, and get some kind of back pay settlement.

Across the table, instead of Earl Schick, they now faced Pete Nash. The differences between the two men were dramatic. Where Schick was cold and aloof, Nash was charming and affable and committed to reaching a settlement that both sides could live with. He was also a consummate negotiator, skilled at moving beyond even the most intractable deadlocks. Charlie McDowell says the union negotiators were immediately struck by the difference:

Schick was the type of person, he had no personality as far as dealing across the table. You were not in the room as far as he was concerned. Schick didn't care a damn if you spoke to him or not. . . . Nash, tall boy, bullshitter, loud tie, you know, big money, but . . . a car sales type, attorney, negotiator. Talk about any subject at any time, didn't care. Joked, laughed, just totally different personality. But he was a shyster, and he was [a] slick trick. I respect him a lot more than I would a person like Schick, because . . . the intent of Schick was to get you out on the street, break your local. That was his goal. Nash's goal was to get a contract.

At two o'clock in the afternoon, Becker opened the negotiations, outlining the list of items the union would need to resolve before an agreement could be reached. The scabs had to go, the union members had to return to their jobs, and all fired workers had to have their records wiped clean. All the lockout-related lawsuits needed to be withdrawn or terminated, including any NLRB actions, and there would have to be some kind of back pay package. The union was strongly opposed to requiring medical examinations for returning members and was committed to bringing back the many union jobs management had contracted out during the lockout. Lastly, Becker told the negotiators, Jack Collins, the pot room committeeman discharged for organizing around safety and health just after RAC took over in 1989, had to be reinstated along with the other terminated workers.

In his opening statement, Pete Nash outlined how the union campaign had brought the company to a state of economic disaster. RAC was due to default on a $180 million loan by June 5, and was also $25 million in debt to creditors and losing $2 million in cash each month. He also threatened that, if the Coca Cola Company's boycott of RAC aluminum went into effect as planned on July 1, the fabrication plant would have to be shut down for the balance of the year.

Nash explained that he had asked for a delay in the NLRB decision because, regardless of what Ries had decided, once it was out, "everybody loses." If the company won, the replacements would stay, the boycott would continue, the union would appeal, and the banks would refuse to extend any additional credit and RAC would bankrupt and close. If the union won, the company would appeal, the union campaign would continue, and the banks would withhold credit. Either way, the plant would have to close.

Nash emphasized, "We want an agreement and a change of attitude in the plant . . . we don't want to revisit who caused the dispute." He said that, although they did "not want to get rid of the scabs, we will talk to you about it . . . we will work at some kind of accommodation." Nash was much less accommodating about back pay. Claiming that "the mere threat of back pay could put this company under," he maintained that RAC simply did not have the money. "Even ten cents on the dollar would sink the company."

Nash was equally adamant about maintaining the new levels of "productivity and flexibility" management had gained with a scab workforce. He argued that RAC was "an entirely different plant than it was eighteen months ago." Since the lockout, they had collapsed twenty-seven job classifications down to six, rewritten job descriptions to include "any task assigned," and contracted out work in janitorial services, environmental technology, mill cleaning, and air conditioning. With this flexibility, the company believed they could get the work done with only twelve hundred workers, five hundred fewer workers and $26 million less in annual payroll than before the lockout. In the same vein, Nash told the union that the company wanted to eliminate company-wide bumping rights by splitting seniority between the fabrication and reduction plants. And, "in order to start fresh," the company wanted to eliminate the precedent-setting value of all past grievance settlements and arbitrations.

The consummate negotiator, Nash said that he understood the union's concerns and that the company "wanted to cut a deal—but it would be irresponsible to cut one that shut down the plant." Nash added that without flexibility they would not have an agreement and without an agreement everything would "go down the tubes." Ominously, he added that, if this was not possible for the union, "we need to know we are not going to get [an agreement]—so we can get on with other plans."

Over the next several weeks the union and the company met almost every day. The company moved on the pot room and pension issues, but held firm on bumping rights between plants and on not bringing back all the terminated workers. Nash also refused to make any promises about not hiring back any of the replacement workers after the locked-out workers returned to their jobs, telling Becker, "If every single scab that's in that plant must never be rehired, then we cannot reach an agreement."

The company also continued to demand much broader work rule changes than the union could swallow. Nash kept pointing out that the language was no different from what the Steelworkers had recently agreed to at ALCOA and other aluminum plants. The local committee members were equally adamant, telling Nash that their members would retire rather than go back to work under the terms he was proposing. Nash harshly replied, "It is the judgment of the company that [if we] leave the flexibility issue to later, the fucking plant just won't be there."

Discussions about the fired workers were equally tough. Fourteen workers had been terminated during the lockout, plus Jack Collins, whose discharge had been upheld by an arbitrator in August 1990, just as the earlier negotiations began. Right off, the company insisted that two of the fired members, Robert Buck II and Gerald Church, Jr., should not be part of the discussions, since Buck was in federal prison and Church was under house arrest for their involvement in the pipe-bombing. The company signaled willingness to reinstate the eight workers who were fired for more minor picket line violations, but refused to consider four of the workers with the most serious charges of picket line violence against them—Bobby Good, Jim Picarella, Eli Morris, and John Morris. As for Jack Collins, the company held firm that his case predated the lockout and so shouldn't even be part of the discussions.

Despite these tensions, both sides continued to talk. Nash repeatedly sought to reassure the union that the new RAC management team was there to bargain in good faith, saying at one point "we are not coming to this table with the same mind set . . . not out to break you or hurt anyone." Becker also kept his committee working, saying we "don't want to quit if there is a prayer" that they could get a settlement both sides could live with.

Driving them forward, and bringing them back to the table each time they felt like walking away, was Judge Ries's impending decision, due to be released the first week of June. Estimates of the back pay liability had reached as high as $120 million, more than $70,000 for each of the seventeen hundred locked-out workers. Both sides knew that, without a settlement acceptable to the rank and file, the company would go into bankruptcy, and the plant would, as Nash kept saying, "go down the tubes." To the chagrin of both Becker and Nash, Ries appeared to be vying for his share of the limelight. He continued talking to the press, repeatedly pointing out that any negotiated settlement would have to be approved by him before the NLRB case could be resolved. It seemed at times that Ries didn't want them to settle.

With the local officers tied up in negotiations in Pittsburgh, Bill Doyle and Clinton Durst were left to hold things together at home. The Tuesday night and Saturday afternoon picnics continued, and negotiators tried hard to attend as often as they could. The biggest event by far that month was the AFL-CIO rally at Fort Unity on Saturday, May 16, attended by more than three thousand locked-out workers, family members, and supporters. Becker, Bowen, Chapman, and the local negotiating team took the weekend off to be present. They were joined by Lynn Williams, United Mine Workers president Richard Trumka, and representatives from the national and state AFL-CIO. With the European tour on hold, Joe Uehlein came to West Virginia as well.

Dressed in their union T-shirts emblazoned with "One day longer" and "We Busted Boyle with Solidarity," the crowd sang the campaign's anthem, "One Day More," along with Dewey Taylor and the Paw Paw Band, as they had so many times before. But this time, with Boyle out and negotiations back on track, that day seemed a lot closer. Richard Trumka, the charismatic young president of the United Mine Workers, told the cheering crowd, "I smell union solidarity and victory at Ravenswood." He continued, "I think this local union is as together as any I have seen."

Returning to Pittsburgh, the negotiators set a self-imposed deadline of May 21, one month after Boyle's forced departure. The next three days they bargained from early in the morning until late at night. They began to hammer out agreements on some of the less contentious issues, but on the bigger issues—the terminations, contracting out, bumping rights, job combinations, back pay, and rehiring the scabs—the two sides seemed as far apart as ever.

With the deadline fast approaching, at 3:30 on the afternoon of May 21 Becker reviewed each of the outstanding issues. They were close to agreement on hiring summer relief workers and decreasing overtime in the pot room, and they seemed to be progressing on a combination of wage increases, cost-of-living adjustments, and profit-sharing, to compensate for giving up the metal price bonus. The company had also backed off on the grievance and arbitration precedents. Everything else, however, remained on the table.

Two hours later the company agreed to the union proposal that, although all employees would be required to undergo physical exams, only "new" medical conditions that occurred after the lockout began could affect their return-to-work status.

While both sides broke for dinner, Becker met with Nash. They talked over back pay. They talked over the return-to-work procedures. They worked out a tentative understanding on summer relief and subcontracting. But with no progress on the terminations and bumping rights, both recognized that no agreement would be reached that night. Just before 10:00 P.M., Becker reported to his team that "maybe we are not going to get there." Talks had broken down.

After a long holiday weekend, both sides returned to Pittsburgh. Nash began by repeating his warnings that RAC was on the verge of bankruptcy, losing several million dollars each month and about to default on its bank loans. The company simply could not afford to spend large amounts of cash up front on a back pay settlement. But since Becker was insistent that they provide some money up front, RAC had reworked their wage proposal, cutting back from fifty cents in the first year down to twenty-five cents, so that they could then afford to offer the locked-out workers $1,000 in back pay.

Becker and the committee were incensed. The company now seemed to be bargaining backwards. Becker told Nash, "We are not interested in any trades like that—maybe we should have not come back to the table." But instead of walking out, they kept talking, with Becker reviewing the union's outstanding issues, one by one. By that evening they had agreements on summer relief and overtime for the pot room and a return of all subcontracted work to the bargaining unit. They had also moved much closer on the back pay and wage package, and the union had won strong successorship language, protecting them if RAC or Rinoman planned to sell the plant.

But the company still refused to move on Collins and the four terminations, and the union held firm on retaining plant-wide bumping rights. They had ironed out most of the details about the return to work, including a commitment that all scabs would be discharged and removed from the plant before the union workers returned, but the company refused to promise not to hire any of the discharged scabs to fill later vacancies. The union bargainers knew that under the National Labor Relations Act there was no duty to bargain, much less reach an agreement, over who management brought in as new-hires. Yet they still tried to impress upon Nash that bringing back any scabs would inflame their members and in the long run, they believed, destroy the plant. Talks continued through the night.

At 4:30 A.M. both sides retreated into their caucuses. More than anything, the remaining terminations stood between the negotiators and settlement. Somehow Nash didn't seem to understand or accept the fact that the union could not, would not, go back unless everyone went back. Becker went one last time to talk privately with Nash. They got right to the point.

I said, "I want to tell you something, Pete. Before we started these negotiations, the company understood that everybody was going to come back to work, and I'm not going to let you piss backwards on me." He said, "George, it's changed." I said "The hell it's changed." And this was on the last night. I said, "Everything else is in place." He said, "We've agreed there's two people," one of them was Picarella and one was somebody else, "that are not coming back." He said, "They're not coming back, and if your committee wants to turn it down, then fine." I said, "Pete, you're misreading something. It isn't that this committee is going to turn anything down. I'm not going let them turn it down. *I'm* telling you that they're both coming back." He said, "Let me make sure I understand you, George. You are telling me that you will sink the deal yourself if they're out." When I nodded, Pete said, "Let me make a phone call."

Two hours later Nash returned. According to Becker, Nash said that it all came down to one worker, Jim Picarella, the union committeeman arrested on weapons charges. Nash told Becker that Picarella wasn't coming back. Becker says he replied,

"Well, it's done, then." [Nash] said, "You mean that everything that we've done, the scabs are fired, the people are coming back, all your folks are coming back, seventeen hundred are coming back, you've got the money. We've worked out an arrangement on back pay, all of this, for the time lost during the strike, some means of trying to compensate. You mean that that's all off?" And I said, "Pete, you're trying to play with me. I'm telling you that if Picarella's out . . . I don't know Picarella, everything that you've said about him may be right. I don't even give a goddamn, it has nothing to do with that at all. It has nothing to do with right or wrong. It has nothing to do with his alleged violence. . . . You talk about violence, there's a lot of different ways to be violent. Some people pick up a gun and commit violence. Other people might want to use a club. But there's no way that what our members may have done out there could even begin to match the violence that you created, that RAC created on the families of those seventeen hundred people for the last twenty months. You destroyed some of those families . . . there's divorces . . . there's people that's committed suicide, people have lost their homes. The chances to educate their children are gone." I said, "There's no way that you can equal that, so if you talk about violence, and maybe it is violence . . . it's all violence. But we're setting all of that aside. I'm telling you they're coming back, and Picarella's coming back, or the discussion ends right now, I'm pulling the committee out now."

Becker says that Nash responded, "Now let me make sure I understand what you're saying. You're telling me that, if we don't get Picarella back, it's all over, right as of this second." Becker responded, "You heard correctly."

[Nash] just looked at me and laughed. He said, "I'll be back." I said, "Pete, I'm giving you fifteen minutes. I'm tired of fucking around with this." He was back in ten minutes. He said, "My principals are very upset. They felt that you took advantage of some things." I said, "Pete, I don't want to listen to this. Is he coming back or isn't he coming back?" He said, "Yes, he's coming back, but everything else stops as of now, there will be no more discussions with you, with your committee. I'm not even empowered to bring my committee face to face with your committee again. There will be no more requests. There will be nothing else. That's done, that's it, that's from my side." I said, "Okay, we got a deal. I'll explain this to our committee."

At 9:15 on the morning of May 27, after bargaining for more than twenty-four hours straight, Becker brought Nash's final offer to the committee. The subcontractors would be gone; all the locked-out workers except Buck (in prison) and Church (under house arrest) would be going back; there would be language on forced overtime and summer relief in the pot rooms; the pension multiplier would be increased by $3.00; COLA would be maintained; wages would increase $1.25 over the life of the agreement; and every locked-out worker would receive $2,000 in back pay. As part of the back pay settlement, a five-year "progress sharing plan" would pay ten percent of RAC's after-tax income into a pool to be distributed to all workers who would have been eligible for back pay under the NLRB decision. Perhaps best of all, all scabs would be fired.

However, the metal price bonus was discontinued, the understanding about job combinations was gone, and the union gave up bumping rights between the plants. Jack Collins, whose discharge had gone to arbitration before the work stoppage began, would not be returning, and, toughest of all, the company made no commitment not to hire back scabs as new employees once all the locked-out workers had returned. The RICO suit was gone, but so was the NLRB back pay, pending Ries's approval of the contract agreement.

It was a tough discussion. But Becker told the committee that this was it. "Understand what I'm saying, there will be no more goddamn meetings. We drew the line. You drew the line on Picarella, everybody had to go back to work. The company has yielded on that. The price of that is the agreement that's laid out here. That means that the few things left on the table that we were still wanting are down the drain. They're done. Now we're playing hardball. Now it's up to you." These were hard words for the committee to hear, but they had gone as far as they could go.

Charlie McDowell was not convinced. He believed that Becker had folded too soon and that Nash was just bluffing about shutting down the plant. "Nash come in originally with orders from Marc Rich to get a contract . . . then the local management started feeding Nash all this bullshit, and the next thing you know . . . it was splitting plants, it was combining all the jobs, and all this bull. . . . And the next thing you know, our primary objective for the union was to get back in the plant. It was stupid. Hell, we had them beat. We had them on their knees. And I think we could have got an honorable contract."

The local's expert on staffing, McDowell worried about the scabs. With retirements, the Steelworkers would not be able to cover all the jobs in the plant. McDowell estimated that more than 150 positions would be vacant on the first day, all of which could be filled with scabs. Stidham and others were equally concerned about the scabs, but they believed they had accomplished everything they could. Most important, the settlement meant that they were all going to return to work and the plant was going to stay open and stay union. Despite their disagreements, the committee agreed to bring the offer back to the members. They would lay out the settlement and answer all questions, but none of them would speak for or against the agreement. The members would decide themselves.

As word leaked back to Ravenswood that a settlement had been reached, workers and their wives gathered at the union hall to hear the latest news. Cindy Butchner, the daughter of a locked-out worker, had purchased 150 red, white, and blue "Victory" bumper stickers and quickly passed them out to the crowd. But the celebrations were muted by the fact that, as one of the workers' wives told a reporter, they were all too "scared to be happy." They were worried about the scabs and they were worried about giving up the back pay from the NLRB. Just that day they had learned that the scabs had been offered a month's salary, three months' benefits, and two weeks' vacation pay to convince them to stay at the plant until the Steelworkers returned. But Marge Flanigan told the *Parkersburg News,* "Back pay was always a pipe dream for the union. We're optimistic. We didn't think they would come back with a contract the men couldn't accept. . . . The Steelworkers had to save them in the 80s. Now, it looks like they need us back to save them in the 90s. Obviously, they've learned the value of a Steelworker because they can't make it with scab labor."

On Saturday, May 30, 1992, locked-out workers gathered at the Charleston Civic Center for a members-only meeting to learn the details of the settlement. For four hours Becker and the committee went over the tentative agreement, point by point, and the realities that had gotten them there. Declaring the settlement a "tremendous victory," Becker emphasized that, although "we didn't win all the issues we wanted to win," the settlement would provide significantly better wages and health and safety provisions than the final offer they were faced with at the onset of the lockout in October 1990. He emphasized that the union had not agreed that the company could offer "preferential hiring" to scabs once the locked-out workers returned, although legally the company was free to hire whomever they pleased. Under no circumstances would the scabs be covered under the union recall procedure. Becker said that both the international and the local committee had done everything they could to urge management not to hire any of the scabs back.

Hard questions were asked and angry speeches were made from the floor, particularly about the job combinations and the scabs. Ironically, James Picarella voiced some of the strongest objections to the settlement, completely unaware of how far Becker had gone to get him back his job.

A ratification ballot was mailed to the homes of the locked-out workers, to be returned no later than June 12. Over the next ten days, the union members debated the settlement. The initial reactions from many were subdued. They had hoped and expected everything to be just like it was before the lockout. Still, they had made significant gains on two of the most important issues, pensions and health and safety in the pot rooms, and the wage package was better than management's final offer eighteen months earlier. And, more than anything, they did not want to turn away from the chance to put the lockout behind them and go back to work.

The hardest thing for most to accept was that some of the scabs might return. As soon as the union meeting was over, management had circulated a memo to salaried workers and scabs reiterating the promise that all permanent replacements would be put on a "preferential hiring list" to fill any positions still vacant after the Steelworkers returned.

Tensions were further exacerbated by the local media coverage. Between May 30 and June 12, as the members considered the proposal and mailed in their ballots, area newspapers were filled with headlines such as "Steelworkers Find Proposal Tough to Take," "History Suggests Union Will Stand Against RAC's Ultimatum," and "Idled Workers Split on Contract's Merits." Reporters seemed to go out of their way to emphasize the most negative aspects of the proposal and to perpetuate false rumors that the union had agreed to "preferential hiring" for scabs. Filled with quotes from replacement workers deriding the union workers and threatening "payback" for the harassment they had received during the lockout, many articles seemed designed specifically to provoke the union workers into rejecting the contract, thereby allowing the scabs to hold on to their jobs. Further fueling the tension was the news that several scabs, including union members who had crossed the picket line such as former local president Gene Richards, had been promoted to supervisory jobs, protecting them from being terminated at the end of the lockout. As the inflammatory coverage escalated, union leaders questioned whether RAC middle managers—many of whom bitterly resented Boyle's ouster and even more bitterly opposed bringing the union workers back—were behind the negative stories.

Still, much of the reporting seemed to fall on deaf ears. Despite misgivings about the scabs, the back pay settlement, and the job combinations, most workers were extremely pleased that the end of the lockout was finally within reach. For those still not convinced, Becker spoke on the lockout hotline:

> Don't be misled by those who want to see this agreement fail. This is quite likely the most important decision you will ever make regarding your future and that of your family and the community. Your union strongly urges you to vote yes. Vote yes to kick the scabs out. Vote yes to return our members and our union to RAC. . . . It is truly a historic settlement worthy of your support. Never in the history of the labor movement has anyone achieved a victory of the magnitude that you have won. Boyle is fired. The scabs are terminated. And you have won your jobs back with full seniority and pension credit to the date of the lockout. This is a stunning victory, not only for organized labor and the membership of Local 5668, but for their families and the community of Ravenswood as well.

But neither Becker nor Nash commented on the other big news circulating about Ravenswood. Starting with a story in the *Parkersburg News* on May 30, the locked-out workers learned that Willy Strothotte had suddenly resigned his position as president of Clarendon Ltd. effective the end of June, severing his business ties to Marc Rich. A spokesperson from Rich's office told reporters that Strothotte had resigned over differences in how Clarendon should be run. For his part Strothotte said very little, telling reporters simply, "I am parting with the Marc Rich group and not with Marc Rich himself. There is nothing sinister about this." Business analysts speculated that Strothotte's departure had something to do with the Ravenswood settlement he had engineered for Rich.

The Steelworkers called Joe Lang in Zug for details. Lang was able to confirm that yes, not only was Strothotte gone, but Rich had forced him out within a few days of Strothotte's giving Nash the go-ahead to allow all the fired workers back. On Friday night, May 29, Strothotte was honored at an awards dinner in Zug. By Monday, Rich had forced him to resign. Lang was convinced that Strothotte's departure was directly related to the events of that last bargaining session. Something had happened that night, serious enough to cause Rich to suddenly sever ties with Strothotte.

Only much later, when George Becker filled in the details of the last night of negotiations, was Lang able to piece together what had happened. In the original restructuring of the RAC board in April, both Craig Davis and Jean Loyer were added to the board. Both men were former aluminum industry CEOs who now worked closely with the Marc Rich group. However, Davis reported directly to Strothotte, while Loyer reported to Rich. Loyer was also a close associate of Ravenswood's new CEO, William Hampshire, both of them having been at one time top officers of Howmet. That final night of negotiations, Nash had checked with Davis and Strothotte prior to settling the contract, reminding Strothotte of their earlier commitment to Becker that all of the locked-out workers would return. Loyer, and by default Marc Rich, were left out of the loop.

When Hampshire and Loyer heard that Picarella and Bobby Good were getting their jobs back, they were furious, both because they had been left out of the decision-making and because they were sympathetic with local management's desire that the fired workers, especially Picarella, never return. That Friday night, May 29, they called Marc Rich. Rich too was furious. "Maybe the worse sin was not checking with him," Becker explained. "Because if he had checked with him, Marc Rich might have agreed with him. And then he'd have been part of it. But the fact that he didn't check with him, then he can be more outraged, he can be more indignant . . . which left Willy Strothotte out in the cold. It was just bing, all of a sudden, Willy Strothotte's name was not to be mentioned by anybody."

In Ravenswood, speculation about Strothotte's departure simply fueled the intense mixture of anticipation and anxiety with which the locked-out workers and their neighbors faced June 12, the day the ratification ballots would be counted. As the day moved closer, the locked-out workers began, for the first time, to think that their ordeal might really end. Slowly the mood began to change from disappointment to celebration. Early on the morning of June 12, six tellers chosen by Dan Stidham from the Local 5668 membership drove to the South Charleston Post Office to pick up the 1,472 ballots. Under the watchful eyes of Joe Chapman and a delegation of reporters, the ballots were taken to the South Charleston Ramada Inn to be counted.

Throughout the morning, Local 5668 members and their families gathered on the grounds of Fort Unity to await the news. By noon they were joined by George Becker, Jim Bowen, and the local negotiating team. The Women's Support Group had made special preparations for the day, supplying André champagne along with the usual beer and soda. On stage, Dewey Taylor and the Paw Paw Band entertained the crowd while they waited for the news. Then, shortly before one o'clock, Joe

Chapman called in the count. Always the showman, Jim Bowen read the results to the crowd. With George Becker and Dan Stidham straining to read over his shoulder, he called out, "And the numbers are, two void ballots, 181 no's, and 1,287 yes votes for the contract." A thunderous cheer rose through the crowd. The agreement had been ratified by an 88 percent margin.

Both champagne and tears flowed freely as Bowen continued, "This story is a story of people with a will to win that in today's conservative atmosphere in America is unheard of, and of support by many other people in the labor movement and out of it that is unparalleled. We owe a debt of gratitude to those who helped us and the working people of America owe a debt of gratitude to the seventeen hundred and one members of Local 5668. . . . We have sent a message to corporate America that working people are not commodities to be cast aside like pieces of worn-out machinery." Bursting with a mixture of relief and elation, George Becker then shouted to the cheering crowd:

> Congratulations to every one of you. I love you and I'd like to hug each one of you! You beat 'em! You beat Emmett R. Boyle! You beat the scabs! I have never been prouder to be a Steelworker, prouder to be a member of organized labor than I am right here today. This local and the community put up with a lot of crap for eighteen months that wasn't necessary, and you won! The labor movement won! When you go back into that plant June 29 with your heads held high, every working person in America will march through the gate with you.

True to character, Dan Stidham was more subdued, but no less proud. He told his members and their families, "We're pretty pleased with the results. We all pulled together and we stayed together. We're better for this whole affair, maybe not monetarily but morally, ethically, and union wise."

Marge Flanigan was ecstatic. She told the group, "How sweet it is. We've waited nineteen months for this. It is a victory. We were determined for justice." The Paw Paw Band then struck up a tune, and Stidham gathered several members of the Women's Support Group in his arms at once and danced around the stage.

Later, Becker, Bowen, Stidham, and the other local leaders drove to the two picket stations and removed the picket signs. The lockout was over.

21

BACK THROUGH THE GATES

At 7:30 A.M. on June 29, 1992, a crowd began to gather in the south parking lot of the Ravenswood plant. Local 5668 members, spouses, children, and union officials from across West Virginia and from Steelworkers headquarters in Pittsburgh joined the five hundred or so workers who would be going in on the first shift. By 8:30 the crowd had swelled to more than two thousand as a parade of workers began to form. You could spot the members of the Women's Support Group, one more time in their bright blue Fort Unity T-shirts. American flags were everywhere.

Emotions bubbled over, from sobbing to defiant chants that broke the dull roar. Choruses of "Solidarity Forever" rippled through the ranks, who stood with arms linked. Machinist Herb Frum could not hold back tears. "This feels great, really great," the thirty-three-year plant veteran said, wiping tears from his eyes. Frum's grandchildren and three daughters had traveled to Ravenswood to see their father return to work. Watching from the sidelines, Frum's daughter Youlanda Crockett told reporters, "This plant has been his life. He could retire, but he'd rather stay in there with his union brothers." She paused to yell, "All right, Dad! Hold your head high" as Frum passed.

The plant entrance seemed strange. The union had removed its pickets and picket shacks some days before and RAC had removed the barbed wire, the bales of hay, and the metal shields that had stood sentry in front of the electrical transformers throughout the lockout. "The Battle of Fort RAC has now ended," Jim Bowen proclaimed.

At the front of the long line of workers, Dan Stidham, George Becker, Joe Chapman, Joe Uehlein, and Jim Bowen took their places behind a banner that read "One Day Longer." For almost two years their work and lives had been intertwined as they had lived through the highs and lows of the campaign. "At various times through this thing," recalls George Becker, "I have to confess, I was really in despair, but I could never ever let anybody know that." But today the doubts had vanished. They were met at the gate by Pete Nash, who had brokered the deal for the return to work.

At the very end of the line was Mike Bailes, the youngest member of the union bargaining team. Just before he went inside the plant, Mike stopped. He turned back and, facing the crowd with a smile, raised his fist high in the air. The crowd roared.

Mike's gesture meant many things to the crowd of workers and their supporters on that glorious June day, as tears flowed and emotions swelled. It stood for their victory. It stood for their perseverance and their defiance against an owner who had seemed unbeatable. But Mike's fist stood for something else that day, something much more personal—his co-worker and close friend, Jimmy Rider. The victory, in some small way, avenged Jimmy's death, when he had collapsed after being forced to work overtime on another warm June day two years before, when the struggle at Ravenswood Aluminum began.

To say that the Steelworkers needed a win at Ravenswood would be an understatement. "It was a hell of a shot in the arm at a time when we desperately needed it," explains George Becker. While the hemorrhaging of basic steel had been stopped and there had been several important victories in the past few years, most notably the win at USX, the union was far from being on stable ground. Particularly since the Steelworkers had drawn a line in the sand and spent millions of dollars on the campaign, defeat was unthinkable. As George Becker had said time and time again, "Failure [was] not an option."

Perhaps the most important consequence of the Ravenswood victory was the restoration of pride and confidence to the Steelworkers union. Even before it was over, the struggle at Ravenswood had become one of those mythic events that are a source of pride for every member of the union. So many Steelworkers had supported the locked-out workers by raising money, handbilling outside beverage companies and sports events, and traveling by car and bus caravans to Fort Unity, that they all shared the sweetness of taking on the toughest of employers and coming out victorious. Lynn Williams suggests,

So much of labor history is what I call romantic defeats, which make great reading but tell terrible, tragic stories about people really being defeated in the end and losing in the end. Our predecessor union was destroyed in 1892 for all practical purposes, and the Steelworkers tried again in 1919 on a national basis and it was destroyed again. And it took John L. Lewis, Philip Murray, the CIO, the Great Depression, Roosevelt, and the Wagner Act for the union to rise. And then, after all the successes, here we are in a period in the eighties where we're under enormous attack. . . . But I think [at Ravenswood] we said to our union, and said to the labor movement and said to the corporate community . . . that when this sort of thing happens, we can fight. We have the ability to fight and to take it on and to have an impact and to change it and to win.

Through the Ravenswood victory the image of the union fundamentally changed in the eyes of employers. The Steelworkers could no longer be seen as a moribund organization, willing to grant concessions if pushed hard enough. "We're not going to walk off the plank, but on the other hand, nobody's going to push us an inch further than we think we have to go. And if we aren't going to do it, we know how to wage a fight," explains District Director Dave Foster.

For a strike threat to be credible, employers must believe that the union is both willing and able to take on an employer and win. After a decade of too many concessions and too many defeats, Ravenswood gave that power back to the Steelworkers. The national and international scope of the campaign and the extensive media coverage ensured that every employer that had a contract with the USWA understood to what lengths the union had been willing to go for a win at Ravenswood, and would be willing to go again.

The Ravenswood victory was not just a triumph for the Steelworkers. It was emblematic of what a newly revived labor movement was becoming. The *New York Times* reported, "In this dowdy, one-stoplight hamlet of 4,200 in the Ohio River Valley, the United Steelworkers of America has shown that a union can still break an employer." If the Steelworkers had the ignominious distinction of losing the strike at Phelps-Dodge, which heralded the worst decade of the American labor movement, they were also responsible for winning the struggle that marked a resurgence for their union and the labor movement nationwide.

While any victory would have been important for the labor movement, the fact that this was a victory over Marc Rich made it much more. Rich's business operations represent a new corporate structure, focused less on direct corporate ownership and more on a widening sphere of control. This complexity and diffuseness makes using traditional methods of leveraging employers considerably more difficult. In the 1920s the labor movement faced the early monopolies, which at first seemed impervious to union pressure. Yet by the late 1930s the Congress of Industrial Organizations was building a new labor movement in these very industries. In the same way, the victory at Ravenswood demonstrated that the new, powerful yet diffuse corporate structures are not impenetrable to workers and their unions.

The unique aspects of the Ravenswood campaign would be difficult to reproduce in other struggles. How often, for example, is the major owner a fugitive from justice? And how often will OSHA find so many safety and health violations? How many unions have a mature and homogeneous workforce steeped in the Mine Worker traditions of solidarity and community? Yet the victory at Ravenswood Aluminum cannot be reduced to its eccentricities. Nor can it be dismissed as an exception, a fluke, or an accident. For the Ravenswood campaign provided a new model for how the American labor movement could win. Against these new kinds of corporations, Ravenswood showed, once and for all, the poverty of the traditional labor strategy of standing on the picket line and waiting for the courts to act. While this strategy may have been effective against stable industrial employers in the 1950s and 1960s in a very different legal and political climate, it was simply not enough to win against the likes of Marc Rich.

Ravenswood also drove home the fact that even the most innovative new tactics alone were insufficient to win in the 1990s. The labor movement had moved forward through the 1980s in fits and starts, experimenting with a variety of new approaches—corporate campaigns, inside tactics, and community support. Yet none had proven the panacea that labor was hoping for. The Ravenswood victory conclusively demonstrated the necessity of using a number of different tactics in sophisticated, multifaceted campaigns. Unlike the struggle against Hormel or International Paper, for example, the Steelworkers did not rely solely on a corporate campaign, but constantly brought pressure to bear in a plethora of new areas. This was not a campaign with just one good idea. New ideas were constantly generated, along with new tactics to try.

There were no magic bullets. Ravenswood worked because the strategies and tactics grew out of careful research, understanding, and continuous analysis and re-analysis of the particulars of the situation at hand. Success depended on being familiar with all the players and their interlocking connections, and diligently tracing the tangled threads of power and money from West Virginia to Washington, to Zug, to Romania, to Venezuela. But success also depended on carefully and creatively selecting where and how to apply the escalating pressure, to make sure to focus the leverage on those individuals and organizations who were both vulnerable to pressure and had the power to end the lockout.

The other important lesson was that Ravenswood was not won by strategy alone. It was also won by the determination, militancy, and solidarity of the members of Local 5668, who held together for twenty long months. Without their will to win and their commitment to hold together, even the best strategy could not have kept them from losing. And this passion for the cause was not restricted to members of Local 5668 and their families. It was something that flowed throughout the Steelworkers organization, all the way up to its leadership, with the unwavering dedication of George Becker and the full support of Lynn Williams.

The efforts to keep local union members involved in the larger strategies allowed the local and the international union to stay united and were fundamental to the process that brought victory. The participation of the members, their voices, their stories, and their commitment, inspired support from workers across the United States and around the globe. From the *Daily News* workers, who virtually adopted Local 5668 throughout the lockout, the twenty thousand workers who booed Marc Rich at the rally in Bucharest, from the unions that donated $25,000 to the union members who gave $25, the locked-out workers touched a chord that only they, not the union staff nor the outside experts, could touch.

This model of inclusion is based on the participation and militancy of local union members in the communities where they live, but it also reaches out to other workers and other communities around the globe. If they fail to tap into this militancy, commitment, and solidarity, American unions will be unable to turn around their more than two-decade slide. The Ravenswood victory shows union members and their leaders how far the commitment and passion of rank-and-file members and their families can go.

Over the last decade we have witnessed important labor struggles that did not make the connections between rank-and-file militancy and participation, strategic pressure, and top union support. Workers showed plenty of local militancy during the strikes at A.E. Staley, Caterpillar, and the Detroit newspapers, but Ravenswood shows that such struggles cannot be won on a local level, no matter how great the commitment and militancy. The Detroit newspapers strike involved six different locals in four international unions, making coordination difficult. But all these fights lacked a comprehensive strategy and a passion to win from the national and regional union leaderships, that could have drawn upon and connected with the local militancy. Without an effective leadership structure and commitment to win, and without a comprehensive, creative, and aggressive escalation strategy, in these struggles and others like them, militancy alone was not enough.

The Ravenswood campaign, however, was about more than just inclusion. The involvement of the members and their families changed the campaign and changed the workers themselves. Leafleting outside the Metals Exchange in London, telling their story to the officials at the NMB Postbank, rallying with their union allies in Prague, listening to Joe Lang testifying before the West Virginia legislature, and sharing food with the caravans of union supporters who traveled to Fort Unity not only were essential to their victory, but forever altered their understanding of their company, their industry, their union, and their world. None of this could have been achieved if they had been just hands for leaflets and bodies for rallies. It was their full participation that grounded and shaped their resolve, and the resolve of their leaders and supporters, to hold out one day longer than Emmett Boyle and Marc Rich.

The workers at Ravenswood were not the only ones who changed and grew during this campaign. The time spent on the picket lines, at the Tuesday night dinners, and working with the Women's Support Group changed the United Steelworkers of America, its officers and staff. In some very fundamental way it reconnected the machinery of the union with its roots on shop floors and in communities across the country.

Even as great as the victory was at Ravenswood, it is important not to exaggerate or mythologize it. Through a combination of strategic brilliance and rank-and-file fortitude, the Steelworkers saved their jobs and their union. But they paid a high price along the road to settlement. Month after month in kitchens in and around Ravenswood, families worried about mortgages, car payments, and college tuition, and how they would survive if the lockout did not end. While some families endured better than others, everyone involved bore the costs. And there were those like Jack Collins and the Rider family, who were the true casualties of the lockout.

Everything the workers risked and lost might never have been put in such jeopardy if the union had not waited until almost five months into the struggle to launch a full-blown coordinated campaign. How differently the story might have unfolded if the research had been done, the membership and community mobilized, and the strategic pressure brought to bear a year before bargaining even started, when Emmett Boyle, and behind him Marc Rich, first took over the plant. The real lessons from Ravenswood are not the globetrotting, the high drama and intrigue, but the careful, thoughtful strategic work that unions need to begin before they ever go out on strike or get locked out.

Even in its finest hour, Ravenswood was still a defensive action by the Steelworkers. The contract they signed was a decent one considering what they had been through and what they stood to lose. Yet the agreement hardly pushed RAC or the aluminum industry to significantly improve the rights and benefits of their workers. These same tactics must be used not just to stave off defeats but to improve the lives of workers and strengthen their unions.

The Steelworkers would hone the strategies developed at Ravenswood and use them again at Bayou Steel in Louisiana, Bridgestone/Firestone,

Inc., and Wheeling-Pittsburgh Steel. Where before they had had to turn to the AFL-CIO's Industrial Union Department and other outside consultants to do the research and conduct much of the campaign, now they worked carefully to build that capacity in-house. But the goal was to do more than train flying squadrons of headquarters staff to rescue strikes and lockouts that had veered out of control. Instead, staff and leaders at every level of the organization were charged with bringing the same kind of research and strategic capacity that the union had used so effectively at Ravenswood to every major contract negotiation the union was involved in, from the very beginning of the campaign.

That Ravenswood was not just a lucky win by the Steelworkers was borne out by what happened to the labor movement in its wake. While not all the victories that followed were directly a product of Ravenswood and the Steelworkers, clearly the model they had pioneered was becoming the approach that labor was using to win across the country. In the heart of downtown Los Angeles, the Service Employees took on the building service industry, organizing ten thousand janitors, primarily Central American immigrants, through an aggressive and creative community-based campaign. In a string of organizing victories in the deep South, the Amalgamated Clothing and Textile Workers (now UNITE) proved that even in low-wage manufacturing, where the most brutal working conditions are coupled with the most virulent employer opposition to organizing, workers will vote union. And, to the amazement of all the pundits, just a few years after the dust settled on the Ravenswood agreement, the United Mine Workers took their members out in a national strike against the Bituminous Coal Operators Association, and won.

But it was the 1997 Teamster victory at UPS that best captured the lessons learned at Ravenswood, proving to the labor movement and the nation that the strike weapon was not lost from labor's arsenal if used as part of a strategic campaign. Just as in Ravenswood, the UPS victory was a product of strategic research, rank-and-file participation, effective internal and external communication, and escalating pressure around the globe. But, unlike Ravenswood, the UPS campaign was launched more than two years before the contract expired, allowing the union to emerge victorious, with overwhelming public support, after a quick, two-week tactical strike.

Labor is on the move again. While unions still have a long way to go to reclaim their past glory, they are once again winning strikes and organizing new workers despite the continued onslaught by employers. Union activists and newly elected officers have reinvigorated labor's involvement in the political process and recommitted to a larger, more inclusive social vision. Even the business press has declared that "Labor's Back."

But it was a victory in a small town in West Virginia that in so many ways set the stage for the changes that followed. Today Ravenswood remains a beacon of hope for American workers and unions who are struggling to re-chart their direction and rebuild a shattered movement. As Richard Trumka told the Ravenswood workers on the eve of their victory,

There was a time when people hadn't heard of Ravenswood. Now Ravenswood is never going to be just another little place in West Virginia. It's going in the labor history books. Ravenswood is the place where working class people said we're going to push until we push you back across the river. It's a place where workers are putting Corporate America on notice that they can't rob us of our dignity.

EPILOGUE

Victory, no matter how great, is never simple. Local 5668, the United Steelworkers of America, the Ravenswood Aluminum Company, and the people who found themselves on either side of the fight would never be the same. Some won, some lost. Some changed while others faltered. Some grew into their humanity, while others just grew more bitter.

After the speeches were made, the celebrations over, and the tears dried, the Steelworkers began the hard work of rebuilding a community, a way of life, and labor relations that were something less than full-scale war. While in some mathematical way much of what happened over the next five years was a result of the lockout and the Steelworkers' victory, most of it was not expected, nor could it have been predicted. It all happened so quickly.

Charlie McDowell never returned to work. Although he had publicly tried to remain neutral about the proposed contract, he couldn't hide his opposition to the agreement and the situation the local would be walking into. "I could not see myself becoming a representative of a damned scab," said McDowell. ". . . I always prided myself that I was representative, regardless if I liked you, hated you, whether you were the youngest, the seniorist. It didn't make no difference to me. Had I stayed . . . there was eighty-four scabs working in the plant, and I was their representative. They notified me at two o'clock. I notified them at five after two that I retired."

McDowell left RAC to work for the state of West Virginia. He was replaced as grievance committee chair by Bud Rose, who ran the Assistance Center during the lockout. More counselor than combatant, Rose was McDowell's polar opposite in both personality and style.

The return to work was hard on everybody. For almost two years Local 5668 members had walked picket lines, distributed food, tracked trucks, traveled across the country and around the world. As much as they had longed for the lockout to be over, and as much as they wanted their jobs back, they had created something wonderful and powerful and they never wanted it to end.

Now, almost two years older, they were back in hot, dirty, factory jobs, sometimes working alongside former scabs. Many workers went back to their old jobs and the old foremen. A significant number, however, were reassigned to new positions, some of which were jobs that had been combined during the lockout. While Pete Nash and Craig Davis welcomed back the members of Local 5668, many of the first-level supervisors felt differently. For eighteen months the supervisors had enjoyed the freedom of being able to do what they wanted without having to conform to a union contract or be subject to a union grievance. Many had built strong ties with the replacement workers. Still other supervisors, including former Local 5668 president Gene Richards, had been scabs themselves, promoted just before the union workers returned to the plant.

In late June 1992, just days before the locked-out workers were due to return to the plant, George Becker, Joe Chapman, and Dan Stidham flew to London to meet with Davis and Nash in a last-ditch effort to keep the company from rehiring the scabs after the locked-out workers returned to the plant. After his first trip to Europe in June 1991, Dan Stidham had vowed he would never fly again. Yet, deeply concerned that the return of the scabs threatened the future of his union, he agreed to make the trip.

After flying all night, the union leaders met Nash and Davis at 6:15 A.M. to argue their case. Davis was polite but dismissive. RAC management was committed to rehiring the scabs to fill any positions left vacant by retiring Steelworkers. They were not going to budge. "It was clear they had made up their minds before we even got there," recalls Dan Stidham. The next day, only thirty-six hours after they departed, Stidham returned alone to Ravenswood, empty-handed. Despite their best efforts, some scabs would be rehired after the union workers went back to work.

Within a few short weeks of the return, the union's worst fears were realized. Almost a hundred scabs were rehired. Both the supervisors and the former scabs did everything they could to antagonize the union workers. Local union officials were concerned they might even provoke a wildcat strike, and worked hard over the first few weeks and months to keep things from boiling over. Somehow the union workers had to find a way to rise above the taunts and harassment and stick with their work. Janice Crawford, who had to work under Gene Richards, explains, "You never forget or forgive really, but you've got to learn to just put it behind you or you go crazy."

It was not an easy transition. "Our families needed counseling. Our men needed counseling. The emotional stress of going back, believe it or not, was worse than them being out," says Marge Flanigan. "They have lost their sense of purpose. They have lost their dedication to the plant. They lost respect for the salaried workers. They need to be programmed to go back in as well as to try to reach some sense of purpose again. That's one of the hardest things that we've had to deal with. It was even more difficult in one respect than the labor dispute."

The issue of the scabs also continued to divide the local. Many in the union felt, like Charlie McDowell, that what the scabs had done was beyond the pale and they could never be forgiven. They wouldn't look the scabs in the eye, utter a word to them, or respond if they were spoken to. Because Ravenswood was a union shop, the scabs automatically became union members after a sixty-day probationary period. Many of the formerly locked-out workers walked out of the union meeting when former scabs were asked to come to the front of the hall to be sworn in.

With more than seven hundred scabs on RAC's hiring list, the situation did not resolve itself quickly. The long list of scabs also prevented sons and daughters of Local 5668 members from finding jobs at the plant, further slowing the healing process. The first year after the settlement saw hundreds of scabs return to work at Ravenswood. Some remained defiant, some eagerly joined the union, many quit. More than 150 scabs who had refused to sign RAC's severance agreement and were never rehired retained their own lawyer and sued the company, for being discharged after being promised permanent positions.

The transition back to "normal life" was made even harder because, not long after the settlement, the local was left without international union staff for several months. For years, Joe Chapman had been asking Jim Bowen to assign him territory closer to his home. On November 15, after two years of working almost full time with Local 5668, Chapman was reassigned to the Huntington office. His replacement, Bill Rose, suffered a heart attack and was unable to devote full attention to Local 5668 until well into 1993.

The settlement was no less difficult for the members of the Women's Support Group. "They threw away the dish cloth and mop and went out to raise hell and to work for justice, and it was hard for them to readjust," remembers Marge Flanigan. "I think the women came very close, some of them, to having a breakdown. In fact, it was a terrible adjustment, just nothing to do all of a sudden. No daily routine, no union hall. We did continue our solidarity picnics, but it was over." As Sue Groves put it, "It was back to housework that we hadn't done in twenty months, and we were not really good going into that. And starting to put the men back on shift work, or whatever routine, and you had to have this meal ready when they came in."

But by late 1993 the company had exhausted the scab hiring list and things began to settle down in Ravenswood and in the local. In fall 1994, in a move that outraged the formerly locked-out workers, the replacement workers who sued the company for being discharged and never returned to work were awarded $16,300 each as part of a settlement with RAC. Today, fewer than three hundred former scabs remain in the plant.

Despite the continued opposition by many workers to the scabs, others realized that, given the advancing age of the workforce, the scabs would need to be organized if the union was truly to be rebuilt. As Woody Call, a young union activist who came from the UMWA, put it, "I have no intention of taking them out for supper, but I know that they're here . . . and there's nothing I can do about it. The only thing I can do about it is try to make union people out of them, the best way that I know how."

Local 5668 emerged from the lockout a different union than the rurally isolated and insular organization that it had been. Dan Stidham reflected,

> It made me a lot more aware of what's going on in the country with the bankruptcies, the buyouts, and the manipulators that manipulate people. It made me a lot more aware of putting the right people in public office, the elective process. . . . At one time I thought we were the only good local union in the Steelworkers. . . the most militant, that we stood up for our rights more than any other local, and I found out that there was a hundred others out there just like us. When they come to your aid, I think the one thing that made me feel the best about this whole thing. . . . It brought us a lot closer together. Naturally we had our differences among ourselves, but no major differences. . . . I think a lot of our members are much more knowledgeable now about what the union's all about, especially the wives. It's made a lot of people aware of what's around them.

After the lockout, Local 5668 became deeply involved in supporting other unions, repaying the generosity they received during their twenty-month ordeal. Their "solidarity fund" donated tens of thousands of dollars to workers and unions all over the country. They also made a point of thanking and recognizing each of their major supporters, inside and outside the union. From Rich Yeselson and Joe Uehlein to Denny Longwell up the river at Ormet, all those who had helped in their struggle were given a silver coin, mined and minted by Steelworkers in Colorado and engraved with the image of the union workers marching back into the plant under the "One Day Longer" banner.

The lockout had trained a cadre of new, younger leaders, many of whom took roles in the local after the settlement. Janice Crawford, a millwright since 1980, became the new treasurer, the first woman officer of the local. Vice-president Bill Doyle retired in poor health and was replaced by Jerry Schoonover. Bill Hendricks, the young safety activist who came out of the pot room, was appointed health and safety chair.

The return to work led to leadership struggles within Local 5668. In the spring of 1994 Dewey Taylor announced his candidacy for president, running against Dan Stidham. Over the twenty months of the lockout, Taylor had emerged as the one of the primary spokespeople for the local, and he enjoyed the limelight, while Stidham's leadership remained behind the scenes. During the campaign, Taylor's team made accusations about financial improprieties at the Assistance Center and about how current union funds were being allocated, especially the solidarity fund. Having brought the local through the lockout and back into the plant, Stidham was too proud to campaign for a position he felt he deserved. Taylor won narrowly and Stidham returned to work in the plant.

After the election, the international investigated all the charges, carefully reviewing the Assistance Center books and the records from the lockout. The investigator found no evidence of any improprieties and gave the local a clean bill of health. In the fall of 1994, seven months after going back into the plant, Dan Stidham retired, after nearly forty years working first for Kaiser and then for RAC, and after almost as many years with the union.

Janice Crawford and Jerry Schoonover were reelected in 1994, but Bud Rose, who had Charlie McDowell's old job as grievance committee chairman, was defeated by Jim Picarella, the last worker negotiated back into the plant and the man whose reinstatement had led to Willy Strothotte's sudden departure from Marc Rich and Co. Picarella's election in particular seemed a return to the more contentious style of Charlie McDowell. With the local still divided, especially over the issue of scabs, members might have felt a more aggressive shop chairman would hold the line. After his defeat, an embittered Bud Rose was offered and took a management position.

Dewey Taylor held the presidency for one term, to be replaced by Jerry Schoonover in 1997. Schoonover had a long history of activism with the local, bridging both the Stidham and Taylor camps. Picarella was defeated in a landslide by Woody Call, who was one of the strongest proponents for healing in the local. Call was thirty-three years old when he came to work in the pot room two weeks before the lockout began. He explains, "I was in the mines for twelve years. . . . I was a union member. I got involved. When I was asked to go out, I went out. Gladly. Now, I became a union man here. This twenty months turned me from a union member to a union man. What I saw, the solidarity, you saw the young and the old and the middle-aged coming together like that. It was unbelievable. I loved it." Call's election in many ways marked a return to the spirit of the lockout. His youth represented a promise for the future of a local where many of the active leaders were near retirement or had already retired.

After some rough months, the Women's Support Group also got back on its feet, providing strike support and assistance for unions in trouble. They still walk picket lines, buy and distribute Christmas presents for children of striking and locked-out workers, and keep in close contact with unions across the country, offering guidance and support gleaned from their own experience. Today, more than six years after the lockout ended, they continue to hold monthly meetings with forty to fifty women attending.

On May 13, 1994, Local 5668 member Jerry Butcher, the driving force behind the construction of Fort Unity, was killed in the plant. He was crushed by a "crucible tilt table," used to pour molten aluminum, that was improperly welded by scabs during the lockout. OSHA fined RAC $1.175 million for the accident. Like bookends, Jimmy Rider's and Jerry Butcher's deaths bracketed the struggle at Ravenswood.

Jim Bowen lost his daughter to an industrial accident in a West Virginia steel mill. Not long after, Bowen became a special assistant to the Steelworkers president, as part of a reorganization in which a number of Steelworker districts, including Bowen's District 23, were merged in a new structure. In 1997, Bowen was elected president of the West Virginia AFL-CIO and Joe Chapman became a member of the state AFL-CIO executive council.

Pete Nash, who had brokered the deal that led to Emmett Boyle's departure and with George Becker hammered out the agreement that brought the union back into the plant, died suddenly in October 1993 of a heart attack. In his *New York Times* obituary, helping to negotiate the settlement at Ravenswood was noted as one of his most important achievements.

George Becker was propelled by the victory into the presidency of the USWA. The tactics pioneered at Ravenswood were used again and again under his leadership—in bitter disputes at Bayou Steel and at Bridgestone/Firestone, where the Steelworkers snatched victory from sure defeat after merging the United Rubber Workers into their union. Through Ravenswood and the victories that followed, Becker rose to national prominence in the American labor movement. In 1996, in a bold move, he announced that the Steelworkers would join with the United Auto Workers and the Machinists to a create a new industrial union, the largest in North America.

The year before, Becker had played a key role in the New Voice campaign, when John Sweeney from SEIU, Richard Trumka from the Mine Workers, and Linda Chavez-Thompson from AFSCME took over the top offices of the AFL-CIO in the first contested election in fifty years. In their platform, the new leaders committed to establishing a new Department of Corporate Affairs designed to support and assist their affiliates in running the kind of strategic bargaining campaigns that brought victory to the Steelworkers at Ravenswood. On October 21, 1995, just before Sweeney gave his inaugural address, Becker led a march and rally on the convention floor honoring and celebrating all those in the labor movement who were locked out or on strike.

The use of aggressive, multifaceted, coordinated campaigns, pioneered at Ravenswood, did not, however, go unchecked. During the Bayou Steel strike, RICO charges were filed against the Steelworkers and were not settled until months after the strike ended. Employers in other industries continue to use RICO as a weapon against strategic campaigns, despite the fact that in almost every case the charges have been settled or dropped before definitive rulings were made by the courts. While no legislation has yet taken shape, conservative members of Congress are pushing to prohibit the use of the very kinds of corporate strategies that brought victory at Ravenswood.

Despite the success of the end-users campaign, business returned quickly to RAC. In December 1994, a new contract was ratified six months in advance of the expiration date. The company continued to hire and, in the fall of 1996, RAC opened its doors in a community celebration, with the cooperation of Local 5668.

Shortly after the settlement, the West Virginia legislature called off its scheduled hearing on Marc Rich. "We don't know how many billionaire fugitives from justice run major corporations in West Virginia from Switzerland. Even if we did come up with legislation, I doubt if the opportunity is out there that often [to use it]," reported committee co-chair Senator Oshel Craigo.

At the same time, the West Virginia Department of Natural Resources called off its investigation of RAC's environmental problems. DNR Director Ed Hamrick reported that the agency had found only low levels of solvent in the spray fields, and in looking for cyanide had found that "all the water samples came back either clean or well below levels of regulatory concern that would cause us to take enforcement action."

Although the environmental lawsuits were dropped, the union later negotiated with the company an Environmental Action Committee that included representatives from the union and the community. Recalls Mike Wright, "We had a settlement, but the health and safety problems were still there, and so were the environmental problems." Two years later, the ground water contamination issue resurfaced when both state and federal environmental protection agencies ordered the company to correct oil pond and cyanide contamination. Admitting "some environmental contamination," this time the company cooperated and agreed to install alternative wastewater spray field technology.

In 1994, Charles Bradley filed a restraining order against his partner in Ormet, Emmett Boyle, to prevent him from purchasing CONALCO, a rolling mill next door to the smelter Ormet owned in Hannibal, Ohio. Bradley alleged in court that "Boyle had used $14.4 million of company funds to pay off his ex-wife's divorce settlement." In a power struggle, Boyle ousted Bradley, whom he referred to as a "disgruntled minority shareholder." Boyle is now the sole director and shareholder of both the Ormet and the CONALCO plants. Although contract talks with the Steelworkers at Ormet went down to the wire in May 1996, the contract was settled without incident. Boyle has not been in touch with Marge Flanigan, although Marge occasionally speaks with Boyle's former wife.

On August 10, 1992, when George Becker was traveling through Switzerland to attend an international conference, Craig Davis invited him to meet with Davis and Marc Rich for lunch in Zug. Becker recounts: "This guy jumps up and comes out, very briskly, and said, 'So this is the famous George Becker,' and he stuck his hand out. And I said, 'And this is the elusive Mr. Marc Rich.' And he busted out laughing." At lunch, the subject of Ravenswood came up. "He said . . . in effect, that it's hard for him not to admire what we were able to do under absolutely impossible odds and that he wanted to thank me. He wanted to thank me, the union, for saving the Ravenswood plant from being forced to close. That it would have been a loss to everybody had we not been able to do what we'd done. . . . He almost spoke of it as an observer, from a third person."

At the end of their conversation, Becker reports, "he said he would welcome me to come back and visit him in Zug when I had more time. And I reciprocated, saying that when he was in the United States, to be sure and stop by and see me. And he laughed. He laughed about that."

Rich's business operations went through dramatic changes in the years after the Ravenswood lockout. Within a year, most of his original partners, Pincus Green, Alexander Hackel, and Felix Posen, retired from his company. In March 1993, in a move that sent shock waves throughout the international finance community, Rich announced that he would be reducing his stake in Marc Rich & Co., A.G., to 15 percent by 1998 and that he was bringing back Willy Strothotte to replace him as chairman and CEO. A year later, Marc Rich and Co., A.G., acquired 100 percent ownership of the Ravenswood plant.

But the changes did not stop there. In August 1994, Rich, who had already reduced his share in Marc Rich & Co., A.G., to 25 percent, changed his company's name to Glencore International, headed by none other than Willy Strothotte. Glencore took over Marc Rich & Co.'s previously undisputed title as the world's largest diversified commodities trading firm. Strothotte told the press the name change was simply an attempt to prevent confusion between Marc Rich & Co., A.G., and Rich's financial and real estate firm, Marc Rich Holding. However, most believed the change was part of an effort to insulate the trading company from the negative press generated by Rich's fugitive status and fomented by the Steelworkers' campaign.

By November 1994 Rich had sold off all his Glencore stock and resigned from the board of directors. RAC spokesperson Pat Gallagher proclaimed, "This is finally a way of being able to say Marc Rich has absolutely no involvement in Ravenswood." Yet, as would be expected, the break with Rich and his associates was neither as clear nor as clean as RAC management might intimate. Within a year Glencore sold the Ravenswood plant to the Century Aluminum Company, which then went public, apparently releasing the plant from the long reach of Rich and Strothotte. But like its predecessor, Clarendon, Ltd., Glencore continues to own 39.6 percent of Century Aluminum's common stock. It operates under a "services" agreement whereby Century purchases primary aluminum and alumina from Glencore and Glencore buys back more than one hundred million pounds of Century primary aluminum products each year. The directors of Century are also no strangers to Marc Rich. Century chairman and CEO Craig Davis helped engineer the Ravenswood settlement and became CEO of RAC. Vice-chairman William Hampshire and Chief Operating Officer Gerald Meyers were presidents of RAC after the lockout. The board also includes Willy Strothotte, chairman and CEO of Glencore International, and Roman Bninski, Century counsel and a former director at the Rich subsidiary Sudelektra AG.

Rich's story doesn't end here. To the astonishment of the financial community, in 1996, nearly three years after handing over the reins to Strothotte, Marc Rich was back in business, with the formation of Marc Rich and Co. Investment, a commodities trading company wholly owned by Marc Rich Holding. To those who questioned why a sixty-one-year-old man would want to reenter the commodities trading business, Rich replied, "I decided to restart trading activity basically to give

young people an opportunity to work in an alternative environment."
Rich was back—or had he ever disappeared?

To this day no action has been taken against Rich by the U.S. Justice
Department. The current FBI International Crime Alert reports that
"his last known address was Himmelrich 28, 63340 Barre, Zug, Switzerland," his residence for years. The report continues, "The U.S. will pay
a reward [up to $750,000] for information that leads to the arrest of
Marc David Rich." In 1994, Rich was granted Israeli citizenship. Like
Switzerland's, Israel's extradition treaty with the United States does not
cover tax evasion.

On September 8, 1996, Rich's daughter Gabrielle died of leukemia
in Seattle. It is not known whether Rich came to this country during
that time.

Six years later Ravenswood is beginning to heal. Only faint traces of
the graffiti painted on barns and roads across Jackson County remain.
Some union members and their families have returned to their churches,
and people are beginning to walk the streets in Ravenswood, Ripley, and
the surrounding towns without looking over their shoulders.

While the aftermath of victory at Ravenswood was complex, something important remains. Despite the scabs, the changes in leadership,
the politics, and the compromises, not all was lost in the aftermath of victory. As Marge Flanigan expresses:

> If this ever happens to us again, we'll be on the front gate, day one. We
> won't be there in cars. We'll be there physically day one at the front gate,
> arm in arm. Take five away, five more of us come in. Just to make a dif-
> ference in labor. Make a difference not [just] for unions, for American
> working people. For others. The government has pushed, forced it on to
> the people, with the low economy and self-esteem, dog-eat-dog. . . . I'd
> like to see that change. I'd like to think that some day we could make a
> difference in it. . . . A labor dispute can almost be the beginning of your
> life, in some ways.

As George Becker reminds us, "Struggles such as this renew the labor
movement. The Ravenswood campaign demonstrated what it takes to win
even where the employer is determined to bust the union—perserverance,
constant escalation of the battle, and a dogged determination not to quit
no matter how bleak the circumstances may look. The labor movement
must be constructive, creative, and ever willing to change but it must
never, never forget how to fight."

NOTES

In all interviews, page numbers refer to the transcripts.

1. Heat Stress, Heat Stroke

1 "was a hundred times harder," interview with Mike Bailes, May 26, 1993, p. 4.

1 "You can't breathe," interview with Mike Schmidt, March 30, 1993, p. 2.

2 "an undetected non-work-related," "Company Files Suit Against Union Official," *Aluminator,* May 3, 1991, p. 3.

2 "they had to do that," interview with Mike Bailes, May 26, 1993, p. 5.

2. Machines in the Garden

4 This account of Washington's Woods is based on Dean W. Moore, *Washington's Woods: A History of Ravenswood and Jackson County, West Virginia* (Parsons, W.Va.: McClain Publishing Co., 1971), pp. 14-27.

4 The description of Henry J. Kaiser is drawn from Mark S. Foster, *Henry J. Kaiser: Builder in the Modern American West* (Austin: University of Texas Press, 1989).

5 The discussion of the aluminum market is drawn from "Skirting Company Town Pitfalls," *Business Week,* September 14, 1957, p. 190.

5 The unemployment problem in Jackson County is referenced in Gerald Sommers, "Labor Recruitment in a Depressed Rural Area," *Monthly Labor Review,* October 1958, p. 1113.

6 "Due to the uncertainty," "Skirting Company Town Pitfalls," p. 194.

6 "The difficulties aren't gone," *ibid.,* p. 195.

6 "When I went to work," interview with Dan Stidham, March 11, 1993, p. 3.

7 "My dad was a vice-president," interview with Bill Doyle, March 29, 1993, p. 1.

7 "We are militant, very militant," interview with Bud Chenoweth, June 8, 1995, p. 12.

7 "The Ohio Valley today," David Reed, "The Ohio Valley—America's Newest Industrial Empire," *Reader's Digest,* December 1963, p. 199.

7 "They were very friendly," interview with Marge Flanigan, March 30, 1993, p. 8.

8 "They had nice household furnishings," *ibid.,* p. 19.

8 "Bobby had a lot," *ibid.,* p. 7.

8 "Every newscast seemed to," Barry Bluestone and Bennett Harrison, *The Deindustrialization of America: Plant Closings, Community Abandonment, and the Dismantling of Basic Industry* (New York: Basic Books, 1982), p. 4.

8 "The number of people," John P. Hoerr, *And the Wolf Finally Came: The Decline of the American Steel Industry* (Pittsburgh: University of Pittsburgh Press, 1988), p. 11.

9 "managing our way," Robert H. Hayes and William J. Abernathy, "Managing Our Way to Economic Decline," *Harvard Business Review,* Vol. 58, 1980, pp. 66–77.

9 The discussion of the Phelps-Dodge strike is based on Jonathan D. Rosenblum, *Copper Crucible: How the Arizona Miners' Strike of 1983 Recast Labor-Management Relations in America* (Ithaca: ILR Press, 1995).

9 "First the DPS," *ibid.*, p. 118.

10 "They wanted to make," interview with Bud Chenoweth, June 8, 1995, p. 2.

10 "I was one of the last," interview with Bud Rose, March 11, 1993, p. 3.

10 "They had purchased," interview with Marge Flanigan, March 30, 1993, p. 10.

3. New Owners, New Management

11 The discussion of the economic decline of the aluminum industry is drawn from "Aluminum Faces a Slow-Growth Future," *Business Week*, October 19, 1981, p. 97.

11 "Penetration in some markets," *ibid.*

11 "Once corporate managers begin," Seymour Melman, *Profits without Production* (New York: Alfred A. Knopf, 1983), p. 20.

11 Phillip Farin and Gary G. Reibsamen, *Aluminum: Profile of an Industry* (New York: McGraw-Hill, 1969).

12 "You . . . have your fixed," "Aluminum Faces . . ."

12 The battle for Kaiser Aluminum is discussed in Todd Mason, Cynthia Green, and Jonathan Levine, "Joe Frates Turns into a Tiger at Kaiser Aluminum," *Business Week*, March 3, 1986, p. 86.

12 "I would rather trust," *ibid.*

13 "It is consistent with our strategy," "Kaiser to Sell Ravenswood Plant," United Press International Newswire, December 13, 1988, p. 1.

13 "quiet, very, very intelligent," interview with Marge Flanigan, March 30, 1993, p. 9.

13 "I enjoyed it," interview with Charlie McDowell, March 30, 1993, p. 22.

14 "It is now time," *Local 5668 Newsletter,* May 1988, pp. 1–3.

14 "Kaiser got rid," interview with Joe Chapman, May 25, 1993, p. 23.

14 "They sent in some," *ibid.*, p. 6.

15 "probably some of the most," interview with John Molovich, April 28, 1993, p. 6.

15 "The company naturally thinks," *ibid.*, p. 8.

15 The discussion of Charles Bradley comes from Bob Regan, "Kaiser to Sell West Virginia Aluminum Units," *American Metal Market*, Vol. 61, No. 241, December 13, 1988, p. 1.

15 "I, along with a lot," United Press International Newswire, December 13, 1988.

16 "They really had," interview with Marge Flanigan, March 2, 1995, p. 10.

17 "Is Emmett Boyle going," interview with Jim Bowen, May 12, 1993, p. 6.

17 "Together, Ravenswood and Ormet," "What Manufacturers Must Do in the '90s to Survive," remarks by R. Emmett Boyle given at the West Virginia Business Summit, The Greenbrier, White Sulphur Springs, W.Va., August 30, 1990.

17 "Ravenswood's future was," "RAC Maps Out Strategy," *Aluminator*, August 22, 1989, p. 1.

17 "This makes it essential," "Chairman's message," *ibid.*, pp. 1–2.

17 "I'll be back," interview with Bill Hendricks, May 26, 1993, pp. 10–11.

17 "Kaiser was a company," interview with Jack Collins, March 2, 1995, pp. 7–8.

18 "the hardest job in the plant," interview with Mike Schmidt, March 30, 1993, p. 2.

18 "When you combine jobs," interview with Jack Collins, March 2, 1995, p. 2.

18 "I've seen people," *ibid.*, pp. 5–6.

19 "And in June, we had," *ibid.*, pp. 8–9.

19 "I looked up there," interview with Bill Hendricks, May 26, 1993, p. 11.

19 "abuse, threats, and retaliation," letter from Jack Collins, September 18, 1989, p. 1.

19 "I believe there is nothing," *ibid.*, p. 2.

19 "was designed, at a minimum," company correspondence dated October 18, 1989.

19 "They felt they could," interview with Jack Collins, March 2, 1995, pp. 20–21.

20 "their first attempt," interview with Charlie McDowell, March 30, 1993, p. 2.

20 "Soon as I got out," interview with Jack Collins, March 2, 1995, p. 17.
20 "I was really friendly," interview with Bill Doyle, March 29, 1993, p. 6.
20 "He put monitors," interview with Dan Stidham, March 11, 1993, p. 9.
21 "They said that we're," interview with Dewey Taylor, March 2, 1995, p. 9.

4. Bargaining

23 "I'm kind of from," interview with Joe Chapman, May 25, 1993, p. 12.
23 "during the tough times," interview with Jim Bowen, May 12, 1993, p. 1.
23 "In 1986 they had," interview with Joe Chapman, May 25, 1993, p. 44.
24 "It was very pointed," *ibid.*, p. 45.
24 "He couldn't stand that," interview with Marge Flanigan, May 27, 1993, p. 11.
25 "Prussian . . . very hierarchical," interview with Paul Whitehead, May 11, 1993, p. 5.
25 "Well, Earl Schick and I," interview with Joe Chapman, May 25, 1993, p. 7.
25 "I just felt threatened," interview with Dewey Taylor, March 31, 1993, p. 2.
25 "I'm not going to," Joe Chapman bargaining notes, October 8, 1990.
25 "there had been a death," *ibid.*
26 "We could see that they," interview with Toby Johnson, May 26, 1993, p. 2.
26 "They come to our," interview with Joe Chapman, May 25, 1993, p. 9.
26 "We saw the orders," anonymous salaried worker, correspondence dated August 2, 1995, p. 1.
26 Toothman's estimate of RAC's expenditures on strike preparations is documented in the National Labor Relations Board transcript dated September 27, 1991, p. 924.
27 "I knew if I got them," interview with Joe Chapman, May 25, 1993, p. 17.

5. Locked Out

29 "We had a meeting," interview with Bill Doyle, March 29, 1993, p. 8.
29 "so it was a hell of a job," *ibid.*
29 "fresh off the street," interview with Bill Hendricks, March 1, 1995, p. 1.
30 "Everybody leaves, everybody," *ibid.*, p. 19.
30 "kind of hem-hawed around," *ibid.*, p. 21.
30 "It was like a big party," interview with an anonymous salaried worker, March 2, 1995, p. 7.
30 "the union has not signed," *ibid.*, p. 8.
31 "You just couldn't believe," *ibid.*
32 "It was dead cold," interview with Bill Hendricks, March 1, 1995, p. 25.
33 "regardless of the status," Michele Carter, "Union Workers Are 'Humiliated' By Company's Actions," *Jackson Herald,* November 3, 1990.
33 "We did not want," interview with Charlie McDowell, March 1, 1995, p. 18.
34 "When someone stands and throws," interview with Joe Chapman, May 25, 1993, p. 26.
34 "war of words and nerves," "RAC Says Issue Is 'Money,'" *Jackson Star News,* November 3, 1990.
34 "unrealistic contract extension proposal," "RAC Lock-out Is On," *Jackson Herald,* November 3, 1990.
35 "It was like a prison," interview with an anonymous salaried worker, March 2, 1995, p. 3.
37 "arrogant black-suited military strangers," Greg Matics, "Judge Orders RAC, Union to Start Talks Here," *Parkersburg Sentinel,* December 5, 1990.
37 "state judge is without authority," "Federal Court Says Fox Can't Force Meetings," *Jackson Star News,* December 15, 1990.
37 "the evidence presented," "State Rules Union Is Not Locked Out," *Jackson Star News,* December 1, 1990.

37 "When they first announced," interview with an anonymous salaried worker, March 2, 1995, p. 22.

37 "if members choose not to," Ron Lewis, "RAC to Retain Replacement Strike Workers," *Parkersburg News*, December 4, 1990.

6. Holding Together

39 "I never felt for a minute," interview with Charlie McDowell, March 1, 1995, p. 35.

39 "There was a pickup truck," interview with Gene Lee, June 26, 1993, p. 1.

40 "We've got to keep," interview with Joe Chapman, May 25, 1993, p. 34.

40 The story of James Watts is drawn from "Caught Between: Cancer Victim Says His Health Led Him Down a Path Which His Heart Didn't Want to Follow," *Jackson Star News*, January 23, 1991, and "Local 5668's Stidham Responds to Watts Story," *Jackson Star News*, January 30, 1991.

41 "people person, well suited," phone interview with Dan Stidham, June 1996.

41 "The International Strike Defense Fund," interview with Dan Stidham, March 11, 1993, pp. 13–14.

42 "We did some checking around," interview with Bud Rose, March 11, 1993, pp. 6–7.

43 "You have to say," interview with Glen Varney, March 30, 1993, p. 16.

43 "It's really heartbreaking," interview with Bud Rose, March 11, 1993, p. 8.

44 "I was real small," interview with Janice Crawford, June 7, 1995, p. 6.

44 "Of course I started," *ibid.*, p. 9.

44 "Well, you've got a mixture," interview with Dan Stidham, March 29, 1993, p. 13.

45 "Well, I'll tell you," interview with Marge Flanigan, March 30, 1993, p. 13.

45 "I had gotten rid," *ibid.*, p. 22.

45 "We met the first time," *ibid.*, p. 5.

46 "[It]. . . gave us something," interview with Flanigan, Groves, and McCoy, March 30, 1993, p. 6.

46 "with babies and banners," see the video "With Babies and Banners," produced by Anne Bohlen, Lyn Goldfarb, and Lorraine Gray, Interface Video Systems, 1986.

46 "we needed positive thinking," "Wives Lend Support," *Local 5668 Lockout Bulletin*, Vol. 5, January 16, 1991, p. 3.

47 "We wanted to let our husbands," Annetta Richardson, "Rally Gives Support to RAC Workers," *Parkersburg News*, January 14, 1991.

47 "I think the workers," *ibid.*

47 "Before unions came along," "Support Pledged," *Jackson Herald*, January 2, 1991.

48 "The support we received today," *ibid.*

7. Civil War

49 "It's like a little old civil war," "ABC News Nightline," September 30, 1991.

49 The discussion of Boyle's letters is drawn from Deborah Harris, Reuters News Service, January 16, 1991, as quoted in *Lockout Bulletin*, Vol. 5, January 16, 1991.

49 "My feeling is that the issue," "Locked Out in America," video production featured on "We Do the Work," Show #308.

49 "Well, you see a strange face," *ibid.*

50 "absolutely made it very plain," interview with Sue Groves, March 30, 1993, p. 17.

50 "I have new Nike shoes," *ibid.*

51 "I lived around here," "ABC News Nightline," Monday, September 30, 1991.

51 "They were all ruffians," interview with Dan Stidham, March 11, 1993, pp. 7–9.

51 "One [of my kids]," interview with Mike Bailes, March 26, 1993, p. 12.

52 "she'd hear the word 'scab,'" *ibid.*, p. 13.

52 "It hurt me that he went in," "Locked Out in America."

52 "I don't know if you can," letter from Gary Cochran to authors, October 25, 1994.

54 The account of company-made jackrocks is based on "Union Member Says New 'Jackrock' Poses Danger," *Jackson Star News*, February 13, 1991.

55 Vester Walker's account is drawn from his written statement given at union hall, January 23, 1991.

55 Marge Flanigan's account is drawn from her written statement given at union hall, February 15, 1991.

55 "I had carpal tunnel surgery," interview with Janice Crawford, March 29, 1993, p. 3.

56 "As soon as you make a bomb," "Feds to Probe Area Bomb Blasts," *Jackson Star News*, March 6, 1991.

57 "The union has stressed," *Lockout Bulletin*, Vol. 5, January 16, 1991, p. 1.

57 "On January 28 we wanted," interview with Marge Flanigan, March 30, 1990, p. 8.

57 "And then on the second," interview with Linda McCoy, March 30, 1993, p. 9.

58 "Don't pay them," interview with Jim Bowen, May 12, 1993, p. 24.

58 "They finally decided," *ibid.*

58 "The idea of having to confront," interview with an anonymous salaried worker, March 2, 1995, p. 21.

58 "Charlie McDowell's vehicle," *ibid.*, p. 20.

58 "The drive-bys were getting bigger," interview with Dan Stidham, March 29, 1993, p. 12.

8. A Second Chance

60 "I don't know whether," "RAC Replacements Now Considered Permanent," *Parkersburg News*, January 10, 1991.

60 "It is our intention," Lois McCann, "Never Return," *Jackson Herald*, January 12, 1991.

60 "If we don't all go back," Greg Matics, "Stidham: If We All Don't Go Back, None of Us Go Back," *Jackson Star News*, January 12, 1991.

61 "Previous offers included deadlines," "Bargaining at a Standstill," *Jackson Herald*, January 19, 1991.

61 "I reached the guy," interview with Paul Whitehead, May 11, 1993, p. 11.

61 "[Rich] said, 'Yeah,'" *ibid.*

62 "This basically clears the air," "Union Workers Drop Charges against Ravenswood Aluminum," *Parkersburg News*, January 19, 1991, and "NLRB's Ferree Says Steelworker Local's Future Could Hinge on an 'Unconditional Offer to Return,'" *Jackson Star News*, January 19, 1991.

62 "we don't think we could," "NLRB's Ferree . . ."

62 "the whole sorry record," "Union Will Consolidate Charges against RAC," *Jackson Star News*, January 23, 1991.

63 "Jesus, you should see that lulu," interview with Joe Chapman, May 25, 1993, p. 40.

63 "And that was like forty-nine pages," *ibid.*

63 "I asked for a suggestion," interview with Richard Brean, May 11, 1993, p. 1.

63 "[Lockouts] are perfectly lawful," *ibid.*, p. 2.

64 "whose main thing was," *ibid.*, p. 12.

65 "I went down," *ibid.*, p. 14.

9. Jump-Start

66 "I built a new home," interview with Joe Chapman, May 25, 1993, p. 45.

67 "I got a call," interview with George Becker, May 10, 1993, p. 5.

67 "We eventually wound up," *ibid.*, p. 6.

67 "Two times up at bat," *ibid.*, pp. 6–7.

68 "I found that our people," *ibid.*, p. 7.
68 "I don't want to talk," *ibid.*, p. 8.
68 "I've been a steelworker," *ibid.*, p. 1.
68 "Well, if you worked," *ibid.*, p. 2.
69 "Like in most things," *ibid.*, p. 7.
70 "Avon Products chairman," "Unions: Labor's New Muscle," *Newsweek*, April 3, 1978, p. 58.
70 "Under the leadership," "Show 'em the Clenched Fist!" *Forbes*, March 20, 1978, p. 31.
70 For an overview of the strike at Hormel, see Hardy Green, *On Strike at Hormel: The Struggle for a Democratic Labor Movement* (Philadelphia: Temple University Press, 1990), and Peter Rachleff, *Hard-Pressed in the Heartland: The Hormel Strike and the Future of the Labor Movement* (Boston: South End Press, 1993).
71 For a discussion of the Morse Cutting Tool strike, see *Labor Research Review*, Vol. 1, No. 1, which is devoted to the Morse strike and victory.
71 For a discussion of the Moog auto plant and "in-plant" strategies, see Jack Metzgar, "Running the Plant Backwards in UAW Region 5," *Labor Research Review*, No. 7, Fall 1985, pp. 35–44.
71 For a discussion of the development of "in-plant" strategies, see *The Inside Game: Winning with Workplace Strategies* (Washington, D.C.: Industrial Union Department, 1986).
73 "pig-in-a-poke," "Union Officials Seem More Open to Profit Plan, But . . ." *Jackson Star News*, January 30, 1991.
73 "Mr. Chapman implies," "RAC Confirms Clarendon as Metal Buyer; Accuses Union of Using Stolen Documents," *Jackson Star News*, January 30, 1991.
73 "We have been aware," *ibid.*

10. *Penetrating the Veil*

75 "tall man with the soft voice," A. Craig Copetas, *Metal Men: Marc Rich and the 10-Billion-Dollar Scam* (New York: G.P. Putnam's Sons, 1985), p. 65.
75 "Marc always felt," *ibid.*
75 "He had the best memory," *ibid.*, p. 67.
75 "Using Philipp Brothers' bank lines," *ibid.*, p. 89.
76 For more on the financing of Marc Rich & Co., A.G., see *ibid.*, p. 85.
76 "Rich had become so big," *ibid.*, p. 115.
76 For an overview of Rich's sale of oil, see *ibid.*, p. 131.
77 "ploy to frustrate," "They Went Thataway: The Strange Case of Marc Rich and Pincus Green: Nineteenth Report by the Committee on Government Operations" (Washington, D.C.: U.S. Government Printing Office, 1992), p. 3.
77 "After Rich fled," "The Sovereign Republic of Marc Rich: America's most wanted white-collar criminal has become a colossus," *Regardie's*, Vol. 10, No. 6, February 1990, p. 46.
77 For more about Rich's academic philanthropy, see Jim Hougan, "King of the World: Marc Rich," *Playboy*, Vol. 41, No. 2, February 1994, p. 104; Shawn Tully, "Why Marc Rich Is Richer than Ever," *Fortune*, August 1, 1988, p. 75; and Universidad Carlos III web site: http://www.uc3m.es/uc3m.
78 "For some time we had been aware," interview with Linda McCoy, March 30, 1993, p. 21.
79 "In other words," interview with Richard Yeselson, April 6, 1993, p. 21.
79 "Through the salaried people," interview with Charlie McDowell, March 30, 1993, p. 7.
80 "an office which seems to be," Bill Hendricks, "Stamford Trip Report," p. 1.

11. *The Campaign Widens*

81 "they were original thinkers," interview with George Becker, May 10, 1993, p. 13.
81 "different ways to screw things up," *ibid.*
81 "tactically doable," interview with Joe Uehlein, April 6, 1993, p. 20.

82 "They'd get there ahead," interview with Bill Doyle, March 29, 1993, p. 10.

82 "Of course, the union has a right," letters from Lynn Williams, April 26, 1991.

82 "As far as we're concerned," "Union Takes Case to RAC Customers," *Parkersburg News*, March 12, 1991.

82 "George Becker brought," interview with Dave Foster, April 18, 1994, p. 1.

83 "We distributed twenty," *ibid.*, p. 2.

83 "It was not a big meeting," *ibid.*, p. 11.

84 "Yeselson [was] sitting there," interview with Richard Brean, May 11, 1993, p. 29.

84 "They knew I was," interview with Richard Yeselson, December 1, 1993, p. 11.

84 "So that was disappointing," *ibid.*, p. 12.

85 "The thing that you could," interview with Richard Yeselson, April 6, 1993, p. 23.

85 "When we looked at this," *ibid.*, p. 20.

85 "USWA Local 5668," Tom Schmitt, "It's 'War,' RAC Sues Union for Millions," *Jackson Star News*, April 10, 1991.

86 "Worlledge talked about letters," "47 Union Members Fingered in Civil Suit Filed by RAC," *Jackson Herald*, April 10, 1991.

86 "On or about January 15," U.S. District Court, Southern District of West Virginia, summons, Case Number A: 91–0401.

86 "By doing so, in my mind," interview with Stuart Israel, December 15, 1993, p. 14.

86 "It shall be unlawful," "Provisions of the Organized Crime Control Act of 1970: Section 100," *RICO Racketeer Influenced and Corrupt Organizations: Business Disputes and the "Racketeering" Laws, Federal and State* (Chicago: Commerce Clearing House, 1984), p. 72.

87 For further discussion of the expansion of RICO, see Deborah Gersh, "Porno Confiscation," *Editor and Publisher*, Vol. 126, August 14, 1993; "Catholic Bishops Sued Under RICO," *Christian Century*, Vol. 111, July 14, 1993, pp. 707–708; "RICO Cuts Both Ways," *Nation*, Vol. 258, February 14, 1994, p. 181; and "Has the Supreme Court Really Turned RICO Upside Down?" *Journal of Criminal Law and Criminology*, Spring 1995, pp. 1223–1257.

87 For a discussion of the increased use of RICO against labor, see David Brody, "Criminalizing the Rights of Labor," *Dissent*, Vol. 42, Summer 1995, pp. 363–367.

87 "I called an attorney," interview with Marge Flanigan, May 2, 1993, p. 7.

88 "The employer did not furnish," Ron Lewis, "RAC Gets Maximum Fine for Dangerous Conditions," *Parkersburg News*, December 14, 1990.

88 "willfully failing," "$27,000 OSHA Fine in RAC Fatal Mishap," *Parkersburg Sentinel*, January 28, 1991.

88 "I am sure the Ravenswood," "USWA Safety Chairman Doyle Speaks Out on RAC Injuries, Questions Stats," *Jackson Star News*, January 30, 1991.

88 "ferret out all the problems," interview with Jim Valenti, January 20, 1994, p. 2.

89 "have no protection whatsoever," "Union Raps RAC Safety," *Jackson Herald*, April 27, 1991.

89 "We consistently log safety statistics," "Charges by Union Arouse RAC Suit," *Parkersburg News*, April 27, 1991.

89 "Lou Albright of the West Virginia," "Remembering . . . With Bowed Heads and Hushed Reverence, the Bell of Death Rang 25 Times," *Jackson Star News*, May 1, 1991.

90 "isn't about raising," *Lockout Bulletin*, Vol. 13, May 8, 1991, p. 1.

90 "If Jimmy were alive," Tobi Elkin, "Organized Labor Rally Draws 5,000 in Jackson," *Parkersburg Sentinel*, April 29, 1991.

12. Small Victories

91 "This was something discussed," Bob Schwarz, "Troubled Plant Buyout," *Charleston Gazette*, May 10, 1991.

91 "They've reshuffled the cards," Ron Lewis, "RAC Chairman to Buy Out Partners," *Parkersburg News*, May 10, 1991.

92 "This has got to end," "Has RAC Harassed?" *Jackson Herald*, May 22, 1991.

92 "Since this complaint appears," Ron Lewis, "Safety Inspectors Denied Access to RAC," *Parkersburg News,* May 24, 1991.

92 "The employee representative," "Ravenswood Turns Back on OSHA Officials," *Parkersburg News,* May 25, 1991.

92 "Our reaction is outrage," "Ravenswood Turns Back on OSHA Officials," *Charleston Gazette,* May 24, 1991.

93 "If any corporation in America," "Ravenswood Turns Back on OSHA Officials," *Parkersburg News,* May 25, 1991.

93 "Wearing 'Fort RAC' shirts," Ron Lewis, "Jay to Investigate RAC'S Rebuff of OSHA Inspectors," *Parkersburg News,* May 30, 1991.

93 "To those who want," "RAC's Ad Claims It's Not the 'Giant,'" *Jackson Star News,* June 1, 1991.

94 "In that area, we fought," interview with George Becker, May 10, 1993, p. 19.

94 "George said to me," interview with Dallas Ellswick, May 29, 1993, p. 10.

95 "What we wanted to do," *ibid.,* pp. 10–15.

95 "[A] lot of them wanted to," *ibid.,* p. 10.

96 "The boy was asleep," interview with Gene Fowler, June 9, 1995, p. 30.

96 "I never did say," interview with Bill Doyle, March 29, 1993, p. 14.

96 "We targeted Budweiser," interview with George Becker, May 10, 1993, p. 20.

97 "We had fifty-pound bags," interview with Bud Rose, May 11, 1993, p. 10.

97 "She would call us," interview with Don Lipscomb, March 29, 1993, pp. 3–4.

98 "We had our own planes," interview with Bill Doyle, March 11, 1993, p. 25.

98 "We started getting a lot of," interview with Glen Varney, March 30, 1993, p. 6.

98 "They got the floor laid," interview with Joe Strickland, March 30, 1993, p. 8.

98 "begging, bumming, and borrowing," *ibid.,* p. 9.

99 "I don't mainly want," interview with Jerry Carpenter, March 31, 1993, p. 3.

99 "We wouldn't have been able," interview with Bud Rose, March 30, 1993, p. 32.

99 "After the drive-bys stopped," interview with Marge Flanigan, March 30, 1993, p. 18.

100 "We soon got sick," *ibid.*

100 "The union realized real quickly," *ibid.*

100 "He was just the type," interview with Marge Flanigan, March 2, 1995, p. 24.

100 "They jumped in," interview with Joe Chapman, May 25, 1993, p. 58.

100 "We may be locked out," "RAC Members Rally around 'Fort Unity,'" *Steelabor,* July/August 1991, p. 14.

100 "Joe Chapman knew that," interview with Dewey Taylor, March 31, 1993, p. 15.

101 "He was up for president," interview with Dan Stidham, March 29, 1993, p. 9.

101 "I set in motion a system," interview with Dallas Ellswick, March 29, 1993, p. 15.

13. Turning toward Europe

102 For more on Rich's investments, see Shawn Tully, "The Lifestyle of Rich, the Infamous," *Fortune,* December 22, 1986, p. 38; Shawn Tully, "Why Marc Rich Is Richer than Ever," *Fortune,* August 1, 1988, p. 75; and Beat Beari and Michael van Orsouw, "Rich, that's me," *Bilanz,* February 1992, pp. 52–57.

102 "He is a titan in the business," Jim Hougan, "King of the World: Marc Rich," *Playboy,* Vol. 41, No. 2, February 1994, p. 104.

102 For more on Rich's corporate structure, see Beari and van Orsouw.

103 "Aluminum Finger," phrase used in Marc Rich interview with Sander Vanocur, "ABC Business World," December 18, 1988.

103 Rich's profit at the Mount Holly plant is reported in Tully, "Why Marc Rich . . ."

103 The estimates of Rich's aluminum capacity are drawn from Beari and van Orsouw.

104 "In metals, it's now Marc Rich," Tully, "Why Marc Rich . . ."

104 For more information on Rich's Jamaica operation, see Craig Copetas, "The Sovereign Republic of Marc Rich," *Regardie's*, Vol. 10, No. 6, February 1990, p. 46.

104 "The cash flow must," *ibid.*

104 "a bargain-basement price," *ibid.*

105 "made a business," Peter Koenig, "Smoking Out Marc Rich," *Institutional Investor*, August 1992, p. 40.

105 Rich's support in the Soviet newspaper is reported in *ibid.* and in Tully, "Why Marc Rich . . ."

105 "No, it couldn't be," Jerry Knight, "With U.S. Subsidy Checks Rolling In, Fugitive Marc Rich Just Gets Richer," *Washington Post*, June 20, 1989, p. D3.

105 "When you call information," *ibid.*

105 "the easiest, cheapest place," interview with Joe Uehlein, December 1, 1993, p. 19.

105 For more on Rich's security apparatus, see Copetas, "The Sovereign Republic . . . " and Hougan, "King of the World."

106 "a rich thug," interview with Michael Boggs, March 31, 1994, p. 2.

106 "In person Rich is," Tully, "The Lifestyle of . . ."

106 Rich's birthday party is reported in Copetas.

106 "Rich apparently flew into England," *ibid.*

107 "I want very badly," Tully, "The Lifestyle of . . ."

107 "a charmer who seemed," Joe Uehlein and Richard Yeselson, "The United Steelworkers of America vs. The Ravenswood Aluminum Company: The Battle of Fort Unity," unpublished manuscript, 1992, pp. 41–42.

108 "I saw a big chalkboard," interview with George Becker, May 10, 1993, p. 34.

108 "The only thing I'm interested in," *ibid.*, pp. 34–35.

108 "Do what you have to," Uehlein and Yeselson, p. 42.

108 "You might want to discuss," fax from Michael Boggs to Marc Rich, May 31, 1991.

109 "Dear Mr. Boggs, Thank you," Marc Rich fax to Michael Boggs, June 6, 1991.

109 "let the dogs loose," interview with Michael Boggs, March 31, 1994, p. 3.

14. In Marc Rich's Backyard

110 "They [SMUV] were consulted," interview with Joe Uehlein, December 1, 1993, p. 3.

111 "had been a Marc Rich watcher," *ibid.*, p. 1.

111 "He was too big to win," interview with Josef Lang, March 29, 1993, p. 7.

112 "The Rich headquarters in Zug," Beat Beari and Michael van Orsouw, "Rich, that's me," *Bilanz*, February 1992, pp. 52–57.

112 "To get beyond that," interview with Joe Uehlein, April 6, 1993, pp. 29–30.

113 "Another secretary picks it up," *ibid.*, p. 30.

113 "Not only was Hergusvil," *ibid.*, p. 31.

114 "Marc Rich will have to," *ibid.*

114 "such things weren't done," Joe Uehlein and Richard Yeselson, "The United Steelworkers of America vs. The Ravenswood Aluminum Corporation: The Battle of Fort Unity," unpublished manuscript, 1992, p. 37.

115 "[D]on't believe those American," interview with Joe Uehlein, April 6, 1993, p. 34.

115 "You've proved it," *ibid.*

115 "The reaction of the rank-and-file," interview with Josef Lang, March 29, 1993, p. 6.

116 "If we did not believe," Andre Marty, "Geht's Um Gewerkschafter, Bleibt Marc Rich Hart Wie Stahl," *Luzerner Neuste Nachrichton*, June 25, 1991, p. 13.

116 "But we won't go away," Dan Stidham statement, USWA press conference in Zug, June 24, 1991.

116 "As broad as a giant," Marty, "Geht's Um Gewerkschafter . . ."

117 "It was the first time," interview with Josef Lang, March 29, 1993, p. 7.

117 "corrupting influence on the Swiss," letter from Lynn Williams to Edouard Brunner, Swiss ambassador, June 26, 1991.

118 "They went in there," interview with Joe Uehlein, December 1, 1993, p. 2.

118 "to take an initiative," Uehlein and Yeselson, p. 39.

118 "I think we all thought," interview with Dewey Taylor, March 2, 1995, pp. 44–45.

119 "How Certain USWA Officials," advertisement, *Jackson Star News,* June 29, 1991, p. A-4.

119 "they might fool someone," *Lockout Bulletin,* Vol. 16, July 8, 1991, p. 1.

119 "I spent my entire working," David Corn, "Workers United, a Town Divided," *Nation,* February 17, 1992, pp. 1, 198.

15. The Tide Turns

120 "Just like that sign," "The Battle of Fort RAC," a video produced by the United Steelworkers, 1991.

120 For more on Rich's deal with the U.S. Mint, see Rose DeWolf, "Union: U.S. Lets Rich Get Richer," *Philadelphia Daily News,* July 12, 1991.

120 For a review of the NLRB complaint, see Mike Jacoby, "NLRB Seeks Back Pay, Re-Hiring of Union," *Parkersburg News,* July 19, 1991.

121 "History Seems to Favor Steel Union," Ron Lewis, *Parkersburg News,* June 23, 1991.

121 "It is the NLRB that will now," "USWA Wins 'Major Battle,' Not War," *Jackson Star News,* July 20, 1991.

121 "I'm very proud of every one," "Chapman Calls for Caution, Cool, Patience," *Jackson Star News,* July 20, 1991.

121 "Remember, it may not happen," Lois McCann, "Victory May Be Close for 'Locked Out' Union Members," *Jackson Herald,* July 20, 1991.

121 "I remember we all went," interview with Richard Brean, May 11, 1993, p. 13.

121 "They restated their commitment," Ron Lewis, "Talks between RAC, Union Break Down," *Parkersburg News,* July 31, 1991.

122 "We really thought that," interview with Richard Brean, May 11, 1993, p. 13.

122 "My thinking went from," interview with Jim Valenti, January 20, 1994, p. 5.

122 "It is common in this type," "OSHA Planning Inspection of Plant," *Aluminator,* August 5, 1991.

123 "I was with one of the," interview with Bill Doyle, March 29, 1993, p. 22.

123 "they got cited for," *ibid.*

124 "I was thinking, well," interview with Jim Hougan, December 16, 1993, p. 4.

124 "Boyle sat across the table," Steelworkers press release, August 14, 1991.

124 "As far as I'm concerned," *ibid.*

124 "If you really put it together," interview with Paul Whitehead, May 11, 1993, p. 49.

125 "My privacy has been violated," R. Emmett Boyle, "Boyle Response to Claims of Union," *Jackson Herald,* August 17, 1991.

125 "I dropped out of the parade," "Local Union Attends D.C. Labor Rally," *Jackson Star News,* September 4, 1991.

126 "We left our mark," *ibid.*

126 "It is RAC's understanding," Ron Lewis, "RAC Not to Blame in Freak Mishap That Killed Va. Man," *Parkersburg News,* September 5, 1991.

126 "It is the right of every," "RAC Sues USWA for Violating Trade Secrets Act," RAC press release, September 9, 1991.

126 "a smoke screen," "Union: RAC Suit a Smoke Screen," *Parkersburg Sentinel,* September 10, 1991.

126 "union members harassed," Larry Cox, "RAC Alleges 35 Incidents of Violence," *Parkersburg Sentinel,* September 11, 1991.

126 "Nothing [RAC] does really," Larry Cox, "RAC Alleges 35 Incidents of Violence," *Parkersburg Sentinel,* September 11, 1991.

127 "[We] believe that," "Stroh's Drops RAC Cans," *Parkersburg Sentinel,* September 13, 1991.

127 "It was a tremendous victory," letter from George Becker to authors, June 22, 1998.

127 "five times smarter," interview with Richard Brean, May 11, 1993, pp. 16–17.

127 "we just started high-fiving," *ibid.,* p. 16.

127 "It's a simple but dignified," *ibid.,* pp. 27–28.

128 "Brean was driven to succeed," interview with Stuart Israel, December 15, 1993, p. 6.

128 "Carol Shore has two little kids," interview with Richard Brean, May 11, 1993, p. 18.

128 "I had thought the case," "Ravenswood Dispute Rages On," *Parkersburg Sentinel,* September 24, 1991.

129 "McDowell came up to me," interview with Richard Brean, May 11, 1993, p. 23.

129 "Sometimes we've made mistakes," interview with Joe Chapman, May 25, 1993, p. 53.

129 "Her [Shore's] direct exam," interview with Richard Brean, May 11, 1993, p. 18.

129 "Joe was kind of the focal point," interview with Stuart Israel, December 15, 1993, p. 9.

129 "I say that because," interview with Richard Brean, May 11, 1993, p. 18.

130 "We got them," *ibid.,* p. 19.

130 "My guess is that," interview with Stuart Israel, December 15, 1993, p. 20.

130 "You see everything here," Jeff Gallatain, "Caravan Gives Idled RAC Workers Wave of Support," *Parkersburg News,* October 27, 1991.

131 "I don't feel like," "RAC Labor Dispute 1 Year Old with No End in Sight," *Parkersburg Sentinel,* October 30, 1991.

16. Escalation

132 "At times I used to," interview with George Becker, May 10, 1993, p. 17.

132 "My response to this," *ibid.*

133 "the corporate campaign was," interview with Lynn Williams, April 4, 1995, p. 6.

133 "The labor movement was in," *ibid.,* pp. 1–2.

133 For a discussion of the Christian Democratic Party, see "CVP in the Canton of Zug: Marc Rich Must Negotiate," *Luzerner Tagblatt,* PMA Clipping Service, Thomas Malionek, translator, June 28, 1991.

133 The Zug parliamentary question is covered in *Interpellation,* Urs Kern, Vorlage Number 7441, June 28, 1991.

133 "to see if they want," John Parry, United Press International, August 3, 1991, dateline Geneva.

133 "in view of the need," International Metalworkers Federation news release, July 4, 1991.

134 "hit a snag," memo from Jan Bom of the FNV to Joe Uehlein, August 1, 1991.

134 "if I tell you," memo from Jim Hougan to George Becker and Joe Uehlein about Rinoman Investment, July 2, 1991.

135 "I know the name," Donald L. Bartlett and James B. Steele, *America: What Went Wrong* (Kansas City: Andrews and McMeel, 1992), p. 104.

135 "Marc Rich, head of," Marc Rich "Wanted" poster, 1991.

136 "We went to the hotel," interview with Jerry Schoonover, March 30, 1993, pp. 10–11.

136 "The back door opened," Joe Uehlein affidavit, prepared December 23, 1991.

137 "A lady came out," *ibid.*

138 "R. Emmett Boyle and," USWA leaflet, October 1991.

138 "Our package consisted of," memo from Rich Yeselson to Joe Uehlein about Vancouver action, October 1, 1991.

139 "When he seen me," interview with Ed Lasko, June 9, 1995, p. 3.

139 "ran the conference," interview with Richard Yeselson, December 1, 1993, p. 26.

140 Donald Bartlett and James B. Steele, "America: What Went Wrong," *Philadelphia Inquirer,* October 23, 1991, p. 1; Michael Shroeder, Maria Mallory, and John Templeman, "Making

Marc Rich Squirm," *Business Week,* November 11, 1991, p. 120; Laura McClure, "Trouble in Ravenswood," *Progressive,* Vol. 28, October 1991, p. 27.

140 "On Thursday, September 19," Pekka Anttila, Tom Lindahl, and Mikko Kiskasaari, "The Police Looking for Marc Rich—Man Who Procures Oil for South Africa Is a NESTE Customer," translation from Finnish magazine *Suera,* Issue 42, October 18, 1991, p. 30.

141 "progressive agenda," letter from Lynn Williams to Vaclav Havel, November 21, 1991.

17. Keeping Up the Pressure

142 "During the past year," "From Inside RAC: A Farewell Letter by a RAC Employee," *Jackson Star News,* November 1, 1991.

143 "Rose Cass came out," interview with the Women's Support Group, June 8, 1995, pp. 1–2.

143 "Christmas 1991 is almost upon us," "An Open Letter to All Concerned and Affected by the Ravenswood Dispute," *Parkersburg News,* December 23, 1991.

144 The RAC press release is covered in "RAC May Be Cited with 275 Violations," *Parkersburg News,* November 19, 1991.

144 "employee exposure to unsecured," "OSHA Proposed $604,500 in Penalties against Ravenswood Aluminum for 231 Alleged Safety and Health Violations," U.S. Department of Labor, Office of Information fax, December 20, 1991.

144 "In this report they," "RAC Receives $600,000 in Fines from OSHA," *Jackson Herald,* December 24, 1991.

144 "Union workers died," *ibid.*

144 "hazards related to open-side," "OSHA Proposed $604,500 . . ."

144 "We have consistently maintained," "RAC Feels Penalties Are Unwarranted," *Jackson Star News,* December 25, 1991.

144 "Just look at the citations," "Union Says Citations Aren't 'Parking Tickets,'" *Jackson Star News,* December 25, 1991.

145 "So you'll see in July," interview with Joe Uehlein, April 6, 1993, p. 3.

145 "In recent years the labor movement," Joe Uehlein and Richard Yeselson, "The United Steelworkers of America vs. The Ravenswood Aluminum Corporation: The Battle of Fort Unity," unpublished manuscript, 1992, p. 60.

145 "We'll give a report," "Stakeholders to Meet for Ravenswood Strike," *Charleston Daily Mail,* January 20, 1992.

145 "George was beside himself," interview with Joe Uehlein, April 6, 1993, p. 40.

146 "After seeing the stakeholders' report," letter from Rich Yeselson to authors, June 11, 1998, p. 3.

146 "To replace striking workers," Paul Nadine, "Ravenswood Workers Rally for End of Dispute," *Charleston Gazette,* January 23, 1992.

146 "Is it fair that," *ibid.*

147 "Mother Jones is fifteen feet tall," interview with Tavia LaFollette, August 1, 1996, p. 3.

147 "like a distorted head," *ibid.*

148 For a discussion of spent potliner, see "Do the Right Thing," Washington State Department of Ecology, November 1990.

148 "they were dumping it outside," interview with Bill Doyle by Laura McClure, April 3, 1991, p. 3.

148 "If one of those barges," *ibid.,* pp. 2–3.

148 "The purpose of the spray fields," "The Ravenswood Aluminum Corp. Sprays Irrigation Fields," report by Disposal Safety Incorporated, October 3, 1991, p. 1.

149 "Ravenswood Aluminum Corp. has been allowed," "State Scrutinizing RAC's Waste Disposal Plans," *Jackson Herald,* January 25, 1992.

149 "They use an oil and water," *ibid.*

149 "The U.S. Environmental Protection Agency," *ibid.*

149 "The chairman recognizes the problems," Ron Lewis, "RAC Called to Charleston Again," *Parkersburg News,* February 12, 1992.

149 "The web is there," Ron Lewis, "Senator Targets Alleged RAC-Fugitive Ties," *Parkersburg News,* February 13, 1992.

149 "While West Virginians have only heard," "Swiss Official Talks—Marc Rich State Senate Topic," *Jackson Herald,* February 12, 1992.

150 "The fact that he denies," *ibid.*

150 "shocking," Ron Lewis, "Senator Targets . . ."

150 "We went out that night," interview with Frank Powers, December 16, 1993, p. 15.

150 "The purpose of today's hearing," "The Strange Case of Marc Rich: Contracting with Tax Fugitives and [sic] at Large in the Alps," hearing before the Government Information, Justice and Agriculture Subcommittee of the Committee on Government Operations, House of Representatives, U.S. Government Printing Office, 1993, p. 105.

151 "The second track is," *ibid.*, pp. 23, 54.

151 "lacked the political will," Jim Hougan, "King of the World; Marc Rich," *Playboy,* Vol. 41, No. 2, February 1994, p. 104.

151 "The Department of Justice has had," "They Went Thataway: The Strange Case of Mark Rich and Pincus Green: Nineteenth Report by the Committee on Government Operations" (U.S. Government Printing Office, 1992), p. 29.

151 "The committee does not understand," *ibid.*, pp. 30–31.

152 "Why Justice should stonewall Congress," Hougan, "King of the World."

152 "if a political decision was made," "They Went Thataway," p. 17.

152 "that neither ORALCO nor," letter from Emmett Boyle to West Virginia Senate, March 3, 1992, p. 1.

152 "We don't go through all this," interview with George Becker, May 10, 1993, p. 22.

153 "The Steelworkers say they want," "Brewers Stop RAC Can Use," *Parkersburg Sentinel,* February 7, 1992.

153 "These regs really meant," interview with Jim Valenti, January 20, 1994, p. 9.

153 "If they've got a big problem," interview with Mike Wright, June 3, 1993, p. 5.

154 "They saw the light," interview with Jim Valenti, January 20, 1994, p. 16.

154 "a farmer, fisherman, swimmer," "Ravenswood to be Sued for Environmental Violations," Steelworkers press release, March 25, 1992.

154 "The river is important," *ibid.*

154 "I am outraged to see," *ibid.*

154 "RAC is using the Ohio River," "Group Accuses RAC of Illegal Dumping," *Parkersburg News,* April 3, 1992.

154 "flaunting the law," *ibid.*

18. Picket Line around the World

156 "short job description," interview with Penny Schantz, April 5, 1994, p. 2.

157 "two giant puppets," press release, Joe Lang, January 24, 1992.

157 "puppetless. I arrived there," interview with Tavia LaFollette, August 1, 1996, p. 4.

158 "American Dream into a nightmare," "Vehement Protests against Marc Rich," *Tages-Anzeiger,* January 29, 1992.

158 "the lever for solving," "U.S. Union on the Trail of Zug Raw Materials Trader," *Berner Zeitung,* January 29, 1992.

158 "Tavia finally returns from," Scott Spencer, "Hope and Hard Times," *Rolling Stone,* April 30, 1992, p. 68.

159 "fallen victim to the flagrant," letter from Willy Strothotte to Agostino Tarabusi, January 27, 1992.

159 "I want to talk to Marc Rich," Scott Spencer, "Hope and Hard Times."

159 "It's time to bring," *ibid.*, pp. 68–69.

160 "So we went in," interview with Dewey Taylor, March 2, 1995, pp. 59–60.

161 "Garment and William Bradford Reynolds," interview with Joe Uehlein, April 6, 1993, p. 59.

161 "to receive a Union delegation," letter from Willy Strothotte delivered to Dewey Taylor and Mike Bailes, January 29, 1992.

162 "the roads are bad," interview with Joe Uehlein, April 6, 1993, p. 60.

162 "rather negotiate than demonstrate," Joe Uehlein and Richard Yeselson, "The United Steelworkers of America vs. The Ravenswood Aluminum Company: The Battle of Fort Unity," unpublished manuscript, 1992, pp. 68–69.

162 "was afraid of getting," interview with George Becker, May 10, 1993, p. 24.

162 "The police here," *ibid.*

163 "After that I got," *ibid.*

165 "The International, I think," interview with Dewey Taylor, March 2, 1995, p. 52.

165 "with Ravenswood workers," "Marc Rich in Romania," *FRATIA News,* February 1992, p. 4.

166 "Well, it was like watching," interview with Charlie McDowell, March 1, 1995, p. 40.

166 "Charlie did a great job," interview with Jerry Fernandez, April 28, 1993, p. 8.

166 "So we go out there," interview with Charlie McDowell, March 1, 1995, p. 43.

167 For a summary of the Czech coverage, see *Lockout Bulletin,* Vol. 17, March 9, 1992, p. 1.

167 "The average person on the street," interview with Dewey Taylor, March 2, 1995, p. 47.

167 "tell the story well," *ibid.*

168 "Huhanantti spun out," Uehlein and Yeselson, p. 140.

168 "We couldn't find an automatic," interview with Dewey Taylor, March 2, 1995, p. 47.

169 "put up a picket line," interview with Bernie Hostein, April 28, 1993, p. 3.

169 "Marc Rich is not," David Corn, "The Search for Marc Rich," *Nation,* February 24, 1992, p. 230.

19. Boyle's Retreat

170 "Hope and Hard Times," Scott Spencer, *Rolling Stone,* April 30, 1992, p. 3.

172 "They wanted to be," interview with George Becker, May 10, 1993, p. 36.

172 "I told him that's crazy," Becker notes from January 7, 1992 meeting with Leon Marcus and Gene Keilen.

173 Garment's friendship with Lane Kirkland is documented in Leonard Garment, *Crazy Rhythm: My Journey from Brooklyn, Jazz, and Wall Street to Nixon's White House, Watergate, and Beyond . . .* (New York: Times Books, 1997), p. 321.

173 "a very impressive guy," interview with George Becker, August 19, 1993, p. 11.

173 "We had a pleasant dinner," interview with George Becker, May 10, 1993, p. 37.

173 "Hey, wait just a," *ibid.*

174 "My message to," *ibid.*

174 "real headway on Ravenswood," *ibid.*

175 "What do you want," *ibid.*, p. 39.

175 "Hey, you had your," *ibid.*

176 "We'll deal with that," interview with George Becker, August 19, 1993, p. 2.

176 "Pete Nash spoke up," interview with George Becker, May 10, 1993, p. 37.

176 "almost settling the future," *ibid.*

176 "had a framework," *ibid.*, p. 38.

176 "We talked about how," *ibid.*, p. 39.

177 "Yes, we kept escalating," interview with George Becker, August 19, 1993, p. 3.

179 "exercise all powers," Verified Complaint for Injunctive Relief and for Relief Pursuant to 8 Del. C.§ 225 (filed in the Court of Chancery of the State of Delaware, April 14, 1992), pp. 21–22.

179 "notorious fugitive from justice," letter from Emmett Boyle to Charles Bradley, Willy Strothotte, and In-Suk Oh, April 13, 1992, p. 2.

179 "on the brink," Verified Complaint . . ., pp. 4–10.

179 "in default to," "RAC on 'Brink of Financial Ruin,'" *Parkersburg News,* May 5, 1992.

180 "Because by April," interview with Joe Uehlein, April 6, 1993, p. 43.

180 "I've completed my decision," "RAC Ruling to be Given in 10 Days," *Parkersburg News,* April 17, 1992.

181 "We understand that," Steelworkers press release, April 17, 1992.

181 "As a result of actions," statement from Emmett Boyle, April 21, 1992.

181 "How does it feel," fax from Jim Bowen to Emmett Boyle, April 21, 1992.

181 "We've been here one," Charles Mason, "Idled Steelworkers Rejoice at Chairman's Ouster," *Parkersburg News,* April 22, 1992, and Lois McCann, "With Boyle Gone, Steelworkers Hopeful," *Jackson Herald,* April 22, 1992.

182 "This is an historic event," "Boyle Out! Talks On!" *Jackson Star News,* April 22, 1992.

182 "We see no reason," Lois McCann, "With Boyle Gone. . ."

182 "We still stand the same," *ibid.*

182 "Which brings us to," RAC speech to first-line managers, April 20, 1992, p. 3.

182 "save at least some," "Information for Current RAC Employees," RAC company memo, April 20, 1992.

183 "I think he's one," Lois McCann, "Worlledge Resigning as RAC Chairman," *Jackson Herald,* April 25, 1992.

183 "issuance of the ALJ's," "Motion to Delay" filed by RAC co-counsel Pete Nash, April 23, 1992.

20. Settlement

185 "Everything we've done," Martha Hodel, "Steelworkers, RAC set for 'Historic Meeting,'" *Parkersburg Sentinel,* April 25, 1992.

186 "Schick was the type," interview with Charlie McDowell, March 30, 1993, p. 15.

187 "We want an agreement," Becker negotiation notes, April 29, 1992.

187 "the mere threat of," *ibid.*

188 "productivity and flexibility," *ibid.*

188 "in order to start fresh," *ibid.*

188 "If every single scab," Becker negotiation notes, April 30, 1992.

188 "It is the judgement," Becker negotiation notes, May 5, 1992.

189 "we are not coming," Becker negotiation notes, May 3, 1992.

190 "I smell union solidarity," Muriel Cooper, "Solidarity works as RAC talks resume," *AFL-CIO News,* May 25, 1992, p. 1.

190 "I think this local," "Huge Crowds Attend AFL-CIO Rally," *Lockout Bulletin,* May 22, 1992, p. 1.

190 "maybe we are not," Chapman negotiation notes, May 21, 1992.

191 "We are not interested," Chapman negotiation notes, May 26, 1992.

191 "I said, 'I want," interview with George Becker, August 19, 1993, p. 6.

192 "'Well, it's done then,'" *ibid.,* pp. 6–7.

193 "Understand what I'm saying," *ibid.,* p. 8.

193 "Nash come in originally," interview with Charlie McDowell, March 30, 1993, p. 17.

194 "scared to be happy," Annetta Richardson, "Union Members Happy, Hopeful," *Parkersburg News,* May 28, 1992.

194 "Back pay was always," *ibid.*

194 "tremendous victory," Ron Lewis, "Union Leaders Remain Neutral on RAC Offer," *Parkersburg News,* May 31, 1992.

195 "Steelworkers Find Proposal Tough," *Charleston Daily Mail,* May 30, 1992; *Parkersburg News,* June 8, 1992, and May 31, 1992.

196 "Don't be misled," "Getting more is just 'baloney.' USWA's Becker says vote yes," *Jackson Star News,* June 10, 1992.

196 "I am parting with," Don Mangan, "Clarendon President Resigns as Union Reaches Strike Settlement," *Fairfield County Business Journal*, June 15, 1991, Sec. 1, p. 1.

197 "Maybe the worse sin," interview with George Becker, August 19, 1993, p. 10.

198 "And the numbers are," "Celebrating! 88% of Steelworkers approve RAC contract," *Jackson Star News*, June 13, 1992.

198 "Congratulations to every one," Muriel Cooper, "Ravenswood Ends 19-month Agony," *AFL-CIO News*, Vol. 37, No. 13, June 22, 1992, p. 1, and "Celebrating! . . ."

198 "We're pretty pleased," "Celebrating! . . ."

198 "How sweet it is," *ibid.*

21. Back through the Gates

199 "This feels great, really great," Annetta Richardson, "2,000 Cheer Union's Return to RAC," *Parkersburg News*, June 30, 1992.

199 "The Battle of Fort RAC," "USWA Will Return to RAC with Unity March," *Jackson Star News*, June 27, 1992.

199 "At various times through," interview with George Becker, August 19, 1993, p. 38.

200 "It was a hell," *ibid.*, p. 39.

200 "So much of labor history," interview with Lynn Williams, April 4, 1995, p. 13.

201 "We're not going to," interview with Dave Foster, April 18, 1994, p. 9.

201 "In this dowdy," Peter T. Kilborn, "Union Shows How to Fight in West Virginia," *New York Times*, May 8, 1992, p. A-12.

206 "Labor's Back," "Big Labor's Long Climb Back," Editorial, *Business Week*, June 24, 1996, p. 170.

206 "There was a time," Muriel Cooper, "Solidarity Works as RAC Talks Resume," *AFL-CIO News*, May 25, 1992, p. 1.

Epilogue

207 "I could not see," interview with Charlie McDowell, March 30, 1993, p. 18.

208 "It was clear they," interview with Dan Stidham, July 29, 1997.

208 "You never forget," interview with Janice Crawford, March 29, 1993, p. 7.

208 "Our families needed counseling," interview with Marge Flanigan, May 2, 1993, p. 17.

209 "They threw away," interview with Marge Flanigan, March 11, 1993, p. 3.

209 "I think the women," interview with Marge Flanigan, March 30, 1993, p. 22.

209 "It was back to," interview with Sue Groves, March 30, 1993, p. 22.

210 "I have no intention," interview with Woody Call, March 30, 1993, p. 7.

210 "It made me a lot," interview with Dan Stidham, March 29, 1993, p. 20.

211 "I was in the mines," interview with Woody Call, March 30, 1993, p. 3.

213 "We don't know how many," Christopher Keough, "Loose Ends Remain after End of RAC Labor Dispute," *Parkersburg News*, July 1, 1992.

213 "all the water samples," "Violation Allegations against RAC Cleared," *Parkersburg Sentinel*, June 4, 1992.

214 "We had a settlement," letter from Mike Wright to authors, June 9, 1998.

214 "some environmental contamination," Century Aluminum 10-K (filed with the Securities and Exchange Commission March 27, 1997), p. 11.

214 "Boyle had used," "Legal Battle over Ormet Corp Settles," *Reuters Financial Service*, August 21, 1995.

214 "disgruntled minority shareholder," *ibid.*

214 "This guy jumps up," interview with George Becker, August 19, 1993, pp. 31–32.

214 "he said he would," *ibid.*, p. 33.

215 For more about Rich's reduction of his interest in Marc Rich & Co., A.G., see Bob Regan, "Marc Rich Owns All Ravenswood," *American Metal Market,* April 13, 1994.

215 Additional information on Glencore can be found in "Glen who?" *Financial Times,* August 3, 1994.

215 "This is finally a way," Ken Ward, Jr., "Rich, Ravenswood Part Company through Sale," *The Charleston Gazette,* December 3, 1994, p. 5A.

215 "I decided to restart," Rachel Carnac, "Rich Return Sets the Markets Buzzing," *Financial Times,* February 9, 1996, p. 27.

216 "his last known address," U.S. Federal Bureau of Investigation, International Crime Alert, U.S. Justice Department web site, http://www.usdoj.gov/criminal/oiafug/fugitive12.htm.

216 "If this ever happens," interview with Marge Flanigan, May 2, 1993, p. 21.

216 "Struggles such as this," letter from George Becker to authors, June 22, 1998.

ACKNOWLEDGMENTS

We feel deeply honored that we were given the opportunity to tell the story of the workers at Ravenswood Aluminum and all who stood with them during their twenty-month lockout. We are profoundly grateful to the officers, staff, and members of Local 5668 and the Steelworkers international who opened up their union, their files, and their hearts to this project. The sixty participants we interviewed, and their willingness to share their stories, made this book possible.

We are especially grateful to George Becker for his unqualified trust and enthusiastic support for the project. From the very beginning, he understood the importance of chronicling labor's stories and learning from our struggles and our victories. He had the vision to understand what it would take to get the job done and the intellect, integrity, and incredible attention to detail to ensure that we got it right. He also had the courage to provide us with unfettered access to every participant and all documents relating to the Ravenswood campaign, and a passion for telling the truth no matter where it led us.

We owe special thanks to Dan Stidham, who welcomed us into his union and his community on each of our many trips to Ravenswood. Dan not only gave us hours of his own time; he encouraged the full participation of his members and their families in the project and provided us access to all the local's records from the lockout.

We want to thank Dewey Taylor, who, upon taking office in 1994, continued to make us welcome in the local and assisted us in collecting photographs and documents and conducting final interviews. Marge Flanigan invited us into her home and provided us a window into the world of the Women's Support Group and the impact of the lockout on Local 5668 members and their families.

Special thanks must be given to Local 5668 member Gene Lee, who entered a massive number of newspaper and magazine articles into a computer database that proved invaluable to us. Gene, along with Dan Stidham, Marge Flanigan, Mike Wright, Joe Uehlein, Rich Yeselson,

Jim English, and George Becker, must be thanked for their careful reading and fact-checking of the final manuscript. Not only did they help enormously with getting all the details straight, they proved to be wise editors and analysts. We give special appreciation to Joe Uehlein for suggesting to the Steelworkers that we chronicle the Ravenswood victory and for working with us and George Becker to make sure the project got off the ground.

Over the years a great number of people have assisted us in this research. Beth Berry, David Litterer, Carolyn Lehan, and Jean Grimes worked hard on the difficult and sometimes tedious job of transcribing the interviews. Assistance was provided by graduate students in the Labor Relations and Research Center at the University of Massachusetts Amherst, including Patrick Crowley, Erin Enwright, Heather Batchelor, Lucinda Kirk-Linn, Andres Astralaga, Dieter Waizenegger, Janine Yodanis, Mark Parker, Ian Boyle, and Katherine Smith, as well as Matthew O'Malley and Rick Farfaglia from Cornell's School of Industrial and Labor Relations. They spent hours sorting files, reviewing transcripts, and conducting library and computer searches. Gary Hubbard and Erin Enwright worked with us on obtaining and selecting photographs. Katie Briggs, a research aide at Cornell ILR, took on as much work as she could to free up Kate to concentrate on writing and editing the manuscript. In the last few months especially, Katie's support and hard work, as well as her genuine enthusiasm for the project, helped ensure that we were able to meet the publication deadlines.

We are especially indebted to Beth Berry of the UMass Labor Center, who, from the preliminary report through each draft, worked tirelessly to edit and prepare the final manuscript. This book benefited not only from her talent, energy, and terrific organization but also from her patience and good humor, without which we would never have survived. We are both better writers, and this is a much better book, thanks to Beth's hard work and commitment to this project.

We also wish to thank Fran Benson and the staff of Cornell University Press. From our very first conversation with her about the Ravenswood project, now more than four years ago, Fran's enthusiasm and support for our efforts helped make this book possible.

Thanks are due also to our many friends, students, colleagues, and family members who read and critiqued earlier drafts of the manuscript. Specifically, we thank Dan Clawson, Jim Rundle, Coert Bonthius, and Marilyn McArthur, as well as the two reviewers for Cornell University Press, for their insights.

Finally, we want to thank our families. Projects like this have a way of imposing on already too busy lives, and we are deeply grateful to our partners and children for picking up the slack at home and for understanding how important this work was to us. This work could have never been completed without their patience, understanding, enthusiasm, and support.

T. J.
K. B.

Index